Space Physics
and
Space Astronomy

MICHAEL D. PAPAGIANNIS

Professor of Physics and Astronomy
Boston University

GORDON AND BREACH SCIENCE PUBLISHERS

New York   Paris   London

To my son
DIMITRIOS

PREFACE

The aim of this book is to introduce the reader to the vast amount of knowledge which we have gained, especially during recent years, in the many areas of Space Physics and Space Astronomy. It is primarily addressed to the advanced undergraduate or beginning graduate students who will soon have to choose their fields of specialization and areas of research. For those who will decide to follow a career in Space Science it will provide a good foundation on which they can build their own area of expertise by further study and research. For those who will choose a different field of Physics or Astronomy for their specialization, it will offer a unique exposure to this exciting scientific frontier, with which every modern Physicist and Astronomer should be acquainted.

This book should also be useful to scientists concentrating in a particular area of space research who would like to read, in a brief and comprehensive way, about a topic in a related field outside their own specialty. With the rapid progress of space research and the development of many related, but still almost independent branches of this field, it is becoming progressively more difficult for any one scientist to remain well acquainted with all these areas. It is also becoming more and more clear, however, that many of the branches of space research are closely interdependent. For this reason, it is not unusual at all when a space radio astronomer starts to read about the ionosphere, an ionospheric physicist about solar x-ray emission, etc. It is hoped that in many such cases this book will serve well as an initial source of information.

Obviously, a book of this kind cannot treat in greath depth the many areas it covers, since any one of the chapters and many of the chapter sections have become books of their own. For this reason, at the end of each chapter the reader is provided with a long list of books that cover the whole chapter or parts of it at greater lentgh. The bibliography of each chapter includes also a list of good review articles, where many more references can be found, and some pioneering papers which have an important educational value.

The first six chapters of the book are devoted to the study of the upper atmosphere, the ionosphere, the magnetosphere, the interplanetary space, the active sun, and the solar-terrestrial relations. These six chapters represent essentially the material covered in a one-semester course on Space Physics which I have given for several years at Boston University. The last two chapters are devoted to the new field of Space Astronomy (Solar and Galactic) which is still in its first steps but holds great promise for the future. A brief review of galactic astronomy is given in Section 8.1 to introduce the unfamiliar readers to the astronomical concepts discussed in Chapter 8. In a one-semester course, these last two chapters can be assigned as reading material, so that the students will become familiar with the wide new horizons which we are now beginning to reach through Space Astronomy. Chapter 7 and 8 should be of especial interest to the increasing number of young scientists and engineers who are now entering this very promising new field.

The book also has two appendices. The first deals with the equation of radiative transfer which is used both in the atmospheres of the planets (Chapter I) and in the atmosphere of the sun (Chapter IV). Appendix II gives a historic review of the landmarks and the past accomplishments of the space era.

In a field that evolves as fast as Space Science it is not possible to write a book that will remain completely up-to-date for many years. Nevertheless, most of the groundwork and the basic principles, which occupy the largest portion of this book, are by now well-established and therefore will not change. There are, however, also several topics that are not yet well understood and where more observations and further theoretical studies are needed. The attention of the reader is drawn to these topics, though I have usually tried to minimize the confusion which results by stating all the alternative or competing theories discussed in the literature.

There is also the element of the new and unexpected discoveries, which is especially prevalent in Space Astronomy. I have tried to convey the excitement of this anticipation in many sections of the book. New developments are the unavoidable shortcoming for any book in a rapidly evolving field. This disadvantage, however, might also have a special educational value, because it has been my experience that students enjoy it very much when their professor tells them about the new exciting things that have happened since the publication of their book.

Boston, Massachusetts   MICHAEL D. PAPAGIANNIS

ACKNOWLEDGMENTS

I would like to thank the following friends and colleagues who read one or more chapters of the book and made many valuable observations and recommendations. In alphabetical order they are: Dr. Jules Aarons of the AFCRL, Dr. Syun-Ichi Akasofu of the University of Alaska, Dr. R.Caro-villano of Boston College, Dr. G. Contopoulos of the University of Thessaloniki, Dr. R. Doyle of the Harvard College Observatory, Dr. G. Fazio of the Smithsonian Astrophysical Observatory, Dr. J. E. Gaustad of the University of California at Berkeley, Dr. G. R. Huguenin of the University of Massachusetts, Dr. S. Krimigis of the John Hopkins Applied Physics Laboratories, and Dr. C. Sagan of Cornell University.

I want to thank also Mrs. M. Matas Smith for the illustrations in the book, Mrs. E. Homer, Miss M. Rajcok, and Miss B. Shatz for typing the draft and final copies of the manuscript, and my graduate students, A. Katz, T. C. Jones, M. Mendillo, H. Mullaney, and F. Strauss, who helped me in many ways in the preparation and proof reading of this book.

M. D. P.

CONTENTS

Preface   vii

Acknowledgments   ix

CHAPTER 1   PLANETARY ATMOSPHERES

1.1   Formation and Evolution of Planetary Atmospheres   .   .   1

1.2   The Structure of the Terrestrial Atmosphere   .   .   .   5

1.3   The Temperature of the Neutral Atmosphere   .   .   .   10

1.4   The Escape of the Atmospheric Gases   14

1.5   The Atmospheres of the Planets   19

1.6   Bibliography   25

CHAPTER 2   THE IONOSPHERE

2.1   Introduction.   28

2.2   The Chapman Layer Theory   30

2.3   The Plasma Frequency   36

2.4   Collision Frequency and Absorption   42

2.5   The Structure of the Ionosphere and the Plasmasphere .   .   46

2.6   Regular and Irregular Variations of the Ionosphere   .   .   53

2.7   Bibliography   61

CHAPTER 3   THE MAGNETOSPHERE

3.1   The Earth's Magnetic Field   63

3.2   The Dipole Magnetic Field   65

3.3   Motions of Charged Particles in a Dipole Magnetic Field   71

3.4   The Radiation Belts   76

3.5   The Boundary and the Tail of the Magnetosphere   .   .   83

3.6   Bibliography   90

CHAPTER 4   THE ACTIVE SUN

4.1   Introduction   92

4.2   The Photosphere   93

4.3   The Chromosphere and the Corona   96

4.4   Sunspots and the Solar Cycle   105

4.5   Faculae, Flares and Prominences   112

4.6   Radio and X-Ray Bursts from the Sun   117

4.7   The Development of an Active Region on the Sun   .   .   123

4.8   Bibliography   126

CHAPTER 5   THE INTERPLANETARY SPACE

5.1   Introduction   128

5.2   Characteristic Parameters of Fully Ionized Plasmas   .   .   129

5.3   Hydrodynamic Equations in the Solar Corona   .   .   .   134

5.4   The Supersonic Flow of the Solar Wind   138

5.5   The Interplanetary Magnetic Field   144

5.6   Interplanetary Dust   150

5.7   Bibliography   157

CHAPTER 6   SOLAR-TERRESTRIAL RELATIONS

6.1   Introduction 158

6.2   Geomagnetic Storms and Ring Currents 161

6.3   Galactic and Solar Cosmic Rays 169

6.4   Auroras 178

6.5   Ionospheric Disturbances 182

6.6   Bibliography 190

CHAPTER 7   SOLAR AND PLANETARY SPACE
ASTRONOMY

7.1   The Domain and the Scope of Space Astronomy . . 192

7.2   Solar X-Ray Astronomy 196

7.3   Ultraviolet, Optical, and Infrared Solar Space Astronomy . 202

7.4   Solar Space Radioastronomy 209

7.5   Planetary Space Astronomy 216

7.6   Bibliography 219

CHAPTER 8   GALACTIC SPACE ASTRONOMY

8.1   Introduction 221

8.2   Gamma-Ray Astronomy 231

8.3   X-Ray Astronomy 237

8.4   Ultraviolet Space Astronomy 246

8.5   Optical and Infrared Space Astronomy 251

8.6   Space Radio Astronomy 258

8.7   Bibliography 265

APPENDIX I   Radiative Transfer and the Eddington Approxi-
             mation   267

APPENDIX II   The Development of the Space Age .   .   .   276

Acknowledgment of Sources   283

Index   285

CHAPTER 1

PLANETARY ATMOSPHERES

1.1 Formation and Evolution of Planetary Atmospheres

Hydrogen and helium are by far the most abundant elements in the universe. For every 1000 atoms of hydrogen there are nearly 100 of helium and only one or two of all the other elements. Most of the helium, it is believed, was formed during the initial explosion (Big Bang) of our expanding universe (see Section 8.1), probably something like 10 billion (10^{10}) years ago. A substantial amount, however, is also produced through the continuous "burning" of hydrogen to helium in the interiors of the stars.

The other elements, which usually are referred to as the heavy elements, are produced in the last fierce stage of nuclear burning of the more massive stars. The active life of these massive stars is of the order of $10^{10}M^{-2.5}$ years, where M is the mass of the star in solar masses, so that a star with $M = 16$ will reach the final phase of its nuclear burning in the relatively short time of approximately 10 million years. These final nuclear reactions, which occur as the temperature in the center of the star rises to about 10^{9}°K, proceed at a very fast rate leading to an explosion as a result of which a substantial part of the mass of the star is ejected into the interstellar space.

In this manner, the galactic matter is continuously enriched in heavy elements and, therefore, newer stars, like our sun, are richer in heavy elements (population I, stars). On the other hand, we can still see less massive stars (their evolution has been slower because of their smaller mass) that were formed from the early galactic material which is very poor in heavy elements (population II, stars).

Our solar system was formed approximately 5 billion (5×10^{9}) years ago from galactic matter already enriched in heavy elements. The most commom of these elements are O, C, N, Ne, Si, Mg, S, Fe, Al, Ca, but they still comprise only about 2% of the total mass of the solar system. The mechanism of the condensation of the initial cloud of dust and gaseous matter to form the sun and planets is still not very well understood. It appears, however, that approximately 90% of this cloud contracted in a central nucleus and formed the *protosun*, whereas the remaining 10%

occupied the space of the present solar system in the form of a diffuse gaseous shell (solar nebula) rotating around the protosun. The rapid rotation forced the nebula to a disc-like shape in the equatorial plane of the sun and induced strong internal turbulent motions. The turbulence of the disc produced gravitational instabilities which resulted in many self-gravitating gaseous condensations. These produced centers of high density which ultimately combined to form the planets of the solar system and most probably their moons too.

A vastly enhanced solar wind in the early stages of the sun's life stripped the inner planets (Mercury, Venus, Earth, and Mars) of all their gaseous matter. What remained was essentially the chemically condensable materials which represented only a small fraction of the total initial mass. The outer planets, on the other hand, were too far to suffer any significant losses at this stage. Thus all of them (Jupiter, Saturn, Uranus, and Neptune) are much more massive than the inner planets and their composition is probably very similar to the composition of the initial nebula from which the solar system was formed. In contrast, the inner planets have a great deficiency in H and He as well as all the other noble gases beyond He. Based on the abundance of silicon, all of which presumably was retained by the earth, the abundances of the noble gases A, Kr, and Xe on our planet are lower by a factor of 10^6–10^7 compared to their respective abundances on the sun.

The formation of the first solid bodies in our solar system took place approximately 4.7×10^9 years ago. This is inferred from the isotopic abundances of radioactive substances in meteorites and in the lunar dust, which show that they have remained essentially in their present solid form for the past 4.5 billion years. Similar studies of rocks and minerals on the surface of the earth show that these minerals have remained unchanged for not more than 3.6×10^9 years. As a result we know very little about the first one billion years of the history of our planet. During this period the earth was formed from the accumulation of smaller solid bodies, lost essentially all of the non-condensable gases of the primitive solar nebula, became heated in the interior due to gravitational contraction and radioactive decay, became chemically differentiated with the heavier elements sinking toward the center, and finally obtained a permanent crust.

The reason why we have not found any rocks older than about 3.6 billion years on earth is not yet entirely clear. The prevailing theory is that the top layers of the land areas were slowly eroded and washed out into the oceans, while the radioactive clocks of the underlying rocks were reset by the enormous pressure from layers above. The ocean floors, on the other hand, are never any older than a few hundred million years because they are continuously replaced by fresh new material from the interior of the earth. This new material flows up in the middle of the oceans along the mid-ocean ridges and starts pushing the oceans floors toward the continents. The ocean floors, together with all the sedimentary deposits, ultimately plunge back into the interior of the earth near the continents forming the deep ocean trenches at the edge of the continents. This continuous motion and replacement of the ocean floors, which by the way is also responsible for the *continental drift*, accounts for their relatively young age. The current studies of lunar rocks and future studies of the atmospheres and rock materials from other planets will, undoubtably, help us to gain a better understanding of the early stages of our solar system.

Chemical analysis of ancient rocks shows that the early atmosphere of the earth was reducing, i.e., it was lacking in free oxygen and contained mainly H_2, CH_4, N_2, NH_3, CN, CO, and H_2O. Part of the water vapor condensed later to form lakes and seas, and at the same time the different sources of energy available such as lightning, volcanic action, solar ultraviolet radiation, etc., acted to form amino-acids and other organic substances from the above-mentioned atmospheric constituents. These chemical reactions have been successfully reproduced in the laboratory by many research groups. The organic compounds were then dissolved in the water bodies on the surface of the earth and formed what is sometimes referred to as the *primordial soup*.

In the depths of the lakes, protected from the hazardous ultraviolet radiation of the sun, these organic chemicals combined through different catalytic reactions (Ponnamperuma, 1968) to form dioxyribonucleic acid (DNA), ribonucleic acid (RNA), and certain proteins called enzymes, which are the beginning of life. This took place on the earth probably close to 3.5 billion years ago. The early organisms resided in a molecular Garden of Eden because they could feed, without doing any work, on the organic substances that were dissolved in the primordial soup. It was not too long, however, before all the available food was consumed (this, in a humorous vein, is often referred to as as the "soup that ate itself"), and the survival of living organisms had to depend on their ability to develop new feeding processes. The crisis was solved by some organisms which managed to start synthesizing their food from H_2O, the CO_2 that was naturally dissolved in the waters, and the energy of the solar rays. This new process, which has oxygen as its by-product, is called *photosynthesis* and it is believed that it appeared on earth approximately 3 billion years ago.

During the pre-paleozoic (pre-camrian) era the process of photosynthesis allowed the accumulation in the atmosphere of small amounts of oxygen,

1*

which in turn gave rise to minute traces of ozone (O_3). Ozone is a very essential element in the evolutionary progress of life because it stops the ultraviolet radiation from the sun before it can reach the ground and cause irreparable damage to all unprotected living organisms. Some oxygen was also produced from the photodissociation of the atmospheric water vapor by the solar ultraviolet radiation and the selective escape of the lighter hydrogen.

In the beginning of the paleozoic era, i.e., about 600 million years ago, the oxygen probably reached a level of about 1 % of its present abundance, and the ozone that was formed allowed life to survive even at very shallow depths of water. This was of great help for the development of life on earth and in a few million years, geologically a very short period of time, the layers of rock which at that time were underwater became full of fossils of many different types of multicellular organisms. Life of many species was now abundant in all oceans, lakes, and rivers around the globe.

In the late Silurian, i.e., about 400 million years ago, oxygen was probably close to 10 % of its present level and ozone had reached a level that permitted the existence of life outside the water. Living organisms again underwent an evolutionary explosion, and by the early Devonian, i.e., about 380 million years ago, great forests appeared on the surface of the earth. This produced a rapid increase in oxygen and, therefore, more protective ozone. Amphibians and insects appeared on the land.

In the more resent history of the earth the levels of CO_2 and O_2 in the atmosphere have been kept approximately constant as a result of the balancing action of photosynthesis and CO_2-binding in carbonic rocks on the one hand, and respiration, burning, oxidation, and volcanic emanations on the other. It is interesting to note that aquatic plants account for at least 4/5 of the presently occurring photosynthesis.

In view of the recent American and Russian missions to Venus which have shown that the atmosphere of Venus is nearly 95 % CO_2 with a ground pressure of about 80–100 Atmospheres, it is of interest to say a few words about the carbon dioxide on our planet. The present level of CO_2 in the terrestrial atmosphere amounts to a total of $\sim 2 \times 10^{18}$ gr of CO_2 of which $\sim 5 \times 10^{17}$ gr are carbon. This is approximately the same or a little higher than the carbon content of the *biomass* (the total mass of all the living organisms on the earth) which, according to the best recent estimates, is of the order of a few time 10^{17} gr. A much larger quantity of CO_2, approximately 60 times the CO_2 content of the atmosphere, is dissolved in the water of the oceans.

It is estimated that the amount of CO_2 that has been released over the ages in the atmosphere of the earth through volcanic action is of the order of 2×10^5 times the present content, i.e., about 4×10^{23} gr. If all this CO_2 had remained in the atmosphere, it would have produced a CO_2 atmosphere similar to the one of Venus with a ground pressure of about 80 Atm. Most of this CO_2, however, combined with metal oxides to form carbonic rocks and minerals such as limestone ($CaCO_3$) and dolomite ($MgCO_3$). A substantial part of it was also taken out from the atmosphere by living organisms and was converted, through photosynthesis, to organic matter with the simultaneous release of free oxygen. An important fraction of this organic matter is continuously withdrawn from the cycle of photosynthesis and oxidation as the remains of dead organisms are mixed with the soil or buried at the bottom of the oceans. If all dead organic matter were burned back into CO_2, it would consume all the available oxygen, and it is only the sedimentation of organic matter that has allowed the accumulation of oxygen in the atmosphere. Unfortunately only a very small fraction ($\sim 3 \times 10^{18}$ gr) of these carbon-containing sediments was transformed into the valuable, concentrated deposits (coal, petroleum, natural gas, etc.) which are known as *fossil fuels.*

If the earth did not have any water, life would have not evolved on our planet to change CO_2 to organic matter and oxygen, and ultimately help to withdraw part of the carbon in the form of organic sediments. Furthermore, there would have been no weathering of the silicate rocks to produce metal oxides which, combined with CO_2, to form carbonic rocks and minerals. As a result, without water the earth would have had a thick atmosphere of CO_2, probably very much like the one of Venus. Due to the closer proximity of Venus to the sun, the photodissociation of water vapor to molecular hydrogen and atomis oxygen was much more effective. The hydrogen escaped into the interplanetary space and the oxygen was probably used to oxidize different carbon compounds such as CH_4 and CO into CO_2. Thus the water supply of Venus was lost instead of condensing to form water basins on the surface of the planet, as it did on earth. It is also possible that because Venus was formed closer to the sun it had less water vapor to start with.

1.2 The Structure of the Terrestrial Atmosphere

To facilitate the study of the atmosphere, we usually divide it into shells with common properties. These shells bear names ending in *sphere* (e.g., stratosphere) and the boundaries between them follow the name of the lower layer with the ending *pause* (e.g., stratopause). The several layers into which the atmosphere is divided vary depending on the principal property of the atmosphere under investigation. One of the most common classifications is when the temperature is used as the guiding parameter. In this case we recognize the following regions of the terrestrial atmosphere:

Troposphere   This is the lowest layer and extends from the ground to about 13 km. The heat source for this region is the surface of the earth, at a temperature of $290 \pm 20°K$ and, therefore, as we move away from the ground the temperature decreases at a rate of $\sim 7°K/km$ reaching a minimum of $210 \pm 20°K$ at the tropopause.

Tropopause   The upper boundary of the troposphere occurring at an altitude of 13 ± 5 km,

Stratosphere   The temperature begins to rise in this region reaching a maximum of $270 \pm 20°K$ at the stratopause. The heating of the stratosphere is due to the absorption of the ultraviolet radiation in the 2000–3000 Å range by the ozone in the ozonosphere. The ozone layer reaches a maximum concentration around 20–25 km.

Stratopause   The upper boundary of the stratosphere occurring at an altitude of 50 ± 5 km.

Mesosphere   The temperature starts decreasing with height in this region due to an energy sink provided by the CO_2 and oxygen emission in the far infrared. It reaches a minimum of $180 \pm 20°K$ at the mesopause.

Mesopause   The upper boundary of the mesosphere occurring at an altitude of 85 ± 5 km.

Thermosphere   The temperature increases steeply with height in this region reaching its peak value of $1500 \pm 500°K$ at the thermopause. The very effective heating source of this layer is the far ultraviolet (100–2000 Å) radiation from the sun which is absorbed in this region causing the photodissociation and photoionization of the atmospheric constituents. Solar particles and meteors also make a small contribution to the heating process.

Thermopause   The upper boundary of the thermosphere occurring at an altitude of 350 ± 100 km. Above this height the atmosphere, due to its high thermal conductivity, maintains the same high temperature (isothermal region) which it first reached at the thermopause.

The physical parameters of an average atmosphere are shown in Figure 1.2-I and are listed in Table 1.2-I. It should be mentioned, however, that the heights of the different layers as well as their corresponding temperatures and densities vary considerably with latitude and are subject to substantial diurnal, seasonal, and solar cycle variations. The diurnal change

TABLE 1.2-I

Altitude in km	Temperature in °K	Density in gr/cm^{-3}	Mean Mol. Weight	Pressure in dyn/cm^2	Mean Free Path in m	Accel. Grav. in cm/s^2
0	288	1.23×10^{-3}	28.96	1.01×10^6	6.63×10^{-8}	981
2	275	1.01×10^{-3}	28.96	7.95×10^5	8.07×10^{-8}	980
4	262	8.19×10^{-4}	28.96	6.17×10^5	9.92×10^{-8}	979
6	249	6.60×10^{-4}	28.96	4.72×10^5	1.23×10^{-7}	979
8	236	5.26×10^{-4}	28.96	3.57×10^5	1.55×10^{-7}	978
10	223	4.14×10^{-4}	28.96	2.65×10^5	1.96×10^{-7}	978
20	217	8.89×10^{-5}	28.96	5.53×10^4	9.14×10^{-7}	975
40	250	4.00×10^{-6}	28.96	2.87×10^3	2.03×10^{-5}	968
60	256	3.06×10^{-7}	28.96	2.25×10^2	2.66×10^{-4}	962
80	181	2.00×10^{-8}	28.96	1.04×10	4.07×10^{-3}	956
100	210	4.97×10^{-10}	28.88	3.01×10^{-1}	1.63×10^{-1}	951
140	714	3.39×10^{-12}	27.20	7.41×10^{-3}	2.25×10	939
180	1156	5.86×10^{-13}	26.15	2.15×10^{-3}	1.25×10^2	927
220	1294	1.99×10^{-13}	24.98	8.58×10^{-4}	3.52×10^2	916
260	1374	8.04×10^{-14}	23.82	3.86×10^{-4}	8.31×10^2	905
300	1432	3.59×10^{-14}	22.66	1.88×10^{-4}	1.77×10^3	894
400	1487	6.50×10^{-15}	19.94	4.03×10^{-5}	8.61×10^3	868
500	1499	1.58×10^{-15}	17.94	1.10×10^{-5}	3.19×10^4	843
600	1506	4.64×10^{-16}	16.84	3.45×10^{-6}	1.02×10^5	819
700	1508	1.54×10^{-16}	16.17	1.19×10^{-6}	2.95×10^5	796

of the atmospheric density, e.g., at different heights is shown in Figure 1.2-II. These results were obtained by computing the drag force which the atmosphere exerts on a satellite and changes, in a readily measurable way, its orbital parameters. It is seen that the density reaches a minimum near 03 : 00 hours local time and a maximum near 14 : 00 hours local time.

The density variations observed represent essentially temperature variations, with the temperature, as in the case of the ground, reaching a maximum in the early afternoon hours. The differences in temperature with local time produce pressure differences around the earth which are compensated by continuous wind motions from the high to the low pressure areas. Deriving, however, from these measurements a quantitative model for the global wind pattern in the thermosphere, is a very difficult theoretical problem. To solve this problem one must build a complex three-dimensional dynamic model for converting densities to pressures. One must also take into account such additional factors as the Coriolis force due to the rotation of the earth, the drag force due to the interaction of the neutral particles with the ionospheric plasma which in turn is coupled with the earth's magnetic field, tidal effects, gravity waves, etc. It has been shown, nevertheless, both theoretically and experimentally that such winds do actually blow almost continuously in the thermosphere. Their velocity is typically of the order of 100 m/s, but under certain conditions such as during magnetic storms it might reach values of the order of 200 m/s or more.

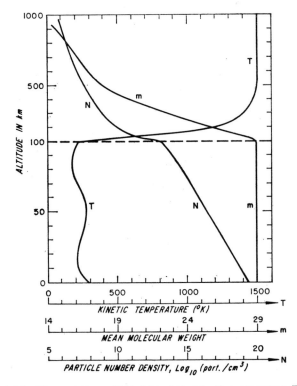

FIGURE 1.2-I   The variation with height of the kinetic temperature T, the mean molecular weight m, and the particle number density N, of the terrestrial atmosphere

As we mentioned earlier, the atmosphere is divided into different layers for different subjects of study. We have already seen the division according to temperature. When our main interest is the chemical composition of the terrestrial atmosphere, we recognize the following regions:

Homosphere   It extends from the ground to about 100 km and is the region where a complete mixing of the atmospheric constituents takes place. The homosphere has a uniform chemical composition with a 28.96 mean molecular weight. It should be noted that layers of minor constituents, such as the ozone layer around 20 km, are nothing more than traces (less than one millionth of the ambient particle density) and, therefore, they do not violate the basic picture of homogeneity prevailing in this region.

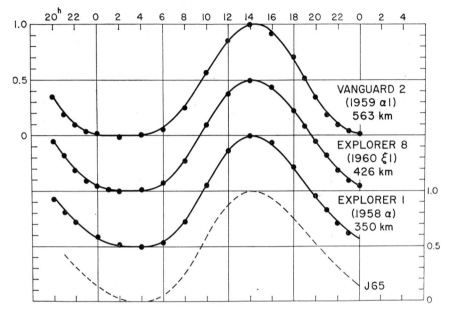

FIGURE 1.2-II   The relative diurnal changes of the atmospheric density at different heights determined from drag measurements with several satellites
(Jaccia, 1970)

Heterosphere   This is the region above 100 km where, due to diffusion and molecular dissociation (e.g., $O_2 \rightarrow O + O$), the chemical composition varies with height and the molecular weight decreases with altitude. Thus around 600 km the average molecular weight is near 16, because the predominant atmospheric constituent is atomic oxygen. Nitrogen dissociates at aslower rate and recombines faster than oxygen so there is very little atomic nitrogen in the upper atmosphere. At even higher altitudes atomic oxygen gives its place to helium and finally helium to hydrogen.

Heliosphere   A layer roughly 1000 km thick between 1000 and 2000 km where helium becomes the main atmospheric constituent.

Protonosphere   The region above about 3000 km where the main constituent is atomic and ionized hydrogen.

Other regions of the upper atmosphere characterized by some common property other than temperature or chemical composition are the following:

Exosphere   It defines the regions from which neutral atoms can escape the gravitational attraction of the earth and extends from approximately 600 km on up.

Ionosphere   This is the region where a partial ionization of the atmospheric constituents takes place. The ionosphere extends from about 70 km on up and reaches a maximum of ionized particle density around 300 km.

Magnetosphere   This is the region where the motion of the ionized particles is governed by the earth's magnetic field. It is rather difficult to define the beginning of the magnetosphere, and one can only roughly place it near 1000 km. The upper limit of the magnetosphere, however, is clearly defined and as expected it is called the *magnetopause*. On the sunlit side of the earth the magnetopause occurs at approximately 10 earth radii, whereas, on the night side of our planet it takes the shape of a long (hundreds of earth radii) cylindrical magnetic tail. The magnetopause defines the boundary of the terrestrial domain beyond which, after a transitional region which is called the *magnetosheath*, starts the vast realm of the interplanetary space.

1.3   The Temperature of the Neutral Atmosphere

Let us first consider the earth as a rapidly rotating solid sphere of radius R. Let the reflectivity of this sphere be such that it reflects a fraction A (*Albedo*) and absorbs the remaining fraction $(1 - A)$ of the incoming solar radiation. Let the sphere also radiate like a black body at an effective temperature T_e. Under conditions of thermal equilibrium (constant temperature) the energy absorbed must be equal to the energy emitted, e.g.,

$$\pi R^2 S_0 (1 - A) = 4\pi R^2 \sigma T_e^4 \qquad (1.3\text{-}1)$$

where S_0 is the solar flux at one Astronomical Unit (1 A.U. $= 1.5 \times 10^{13}$ cm, is the average distance of the earth from the sun), and σ is the Stefan-Boltzmann constant. The flux S_0 is usually called the *solar constant* and can be computed from the values of the solar radius R_s, the Astronomical Unit, and the effective temperature of the sun T_s,

$$S_0 = \sigma T_s^4 \left(\frac{R_s}{1 \text{ A. U.}} \right)^2 \qquad (1.3\text{-}2)$$

The value of the solar constant is $S_0 \simeq 1.4 \times 10^6$ erg cm^{-2} s^{-1} $\simeq 2$ cal cm^{-2} min^{-1}. The albedo of the earth is $A \simeq 0.4$. Thus, combining

(1.3-1) and (1.3-2), we obtain,

$$T_e = T_s \left(\frac{R_s}{1 \text{ A. U.}} \right)^{1/2} \left(\frac{1 - A}{4} \right)^{1/4}$$

$$= 5,800 \left(\frac{6.9 \times 10^{10}}{1.5 \times 10^{13}} \right)^{1/2} (0.15)^{1/4} \simeq 245°K \qquad (1.3-3)$$

The temperature of (1.3-3) is approximately 45°K lower than the average ground temperature T_g of the earth. The difference is due to the *greenhouse effect* of the terrestrial atmosphere which acts as follows. The incident solar radiation has its maximum intensity in the visible portion of the spectrum and, therefore, passes with practically no attentuation through the transparent atmosphere of the earth. Thus the $(1 - A)$ fraction of the solar radiation that is not reflected back, is absorbed by the ground and heats it up. The earth radiates as a black body at a temperature $T_g \simeq 290°K$, which is the average temperature of its surface. At $T_g \simeq 290°K$ most of the emitted energy is in the infrared region (the maximum intensity, according to Wien's law, occurs at a wavelength $\lambda_m \simeq 0.29/290 \simeq 10^{-3}$ cm $\simeq 10\,\mu$). The infrared spectrum, however, is strongly absorbed by the triatomic molecules of the atmosphere, namely CO_2, H_2O, and O_3. The energy absorbed by these molecules is reemitted in part toward the outer space and in part toward the ground, thus providing an additional heating source for the surface of the earth.

For the ground to remain at a constant temperature (thermal equilibrium), the upward flux from the ground σT_g^4 must be equal to the downward flux from the triatomic molecules of the atmosphere, which let us call F_d, plus the flux of solar radiation absorbed by the earth which, as seen from (1.3-1), is equal to σT_e^4 (see Figure 1.3-I). Thus we must have,

$$\sigma T_g^4 = \sigma T_e^4 + F_d \qquad (1.3-4)$$

To compute F_d we must solve the equation which describes the passage, partial absorption, and emission of infrared radiation through the terrestrial atmosphere. This is the equation of radiative transfer (see Appendix I) which, in its full generality, can be extremely complex (Sagan, 1969). Under certain simplifying assumptions, however, such as local thermodynamic equilibrium (LTE), radiative equilibrium, absorption independent of frequency (grey atmosphere), etc., the problem can be solved analytically In the simplest case the solution of the equation of radiative transfer is given by the *Eddington approximation*, which is derived in Appendix I. As seen from equation (A-28) of Appendix I, the intensity of the downward flowing radiation is,

$$I_d = \frac{F}{\pi}\left(\frac{3}{4}\tau\right) \qquad (1.3\text{-}5)$$

and since from (A-7) we have that $F_d = \pi I_d$, it follows that,

$$F_d = \pi I_d = F\left(\frac{3}{4}\tau\right) \qquad (1.3\text{-}6)$$

But, as seen from (A-18), $F = \sigma T_e^4$ and therefore we have,

$$F_d = F\frac{3}{4}\tau_0 = \sigma T_e^4 \frac{3}{4}\tau_0 \qquad (1.3\text{-}7)$$

where τ_0 is the opacity in the infrared of the terrestrial atmosphere. Using (1.3-7) we can now write (1.3-4) in the form,

$$\sigma T_g^4 = \sigma T_e^4 + \sigma T_e^4\left(\frac{3}{4}\tau_0\right) = \sigma T_e^4\left(1 + \frac{3}{4}\tau_0\right) \qquad (1.3\text{-}8)$$

which relates the ground temperature T_g to the effective temperature T_e and the opacity τ_0. The resulting relation is,

$$T_g = T_e\left(1 + \frac{3}{4}\tau_0\right)^{1/4} \qquad (1.3\text{-}9)$$

It has been observed that approximately 85% of the infrared radiation is absorbed in the atmosphere and only 15% of the ground intensity (I_g) makes it through the earth's atmosphere. Since to a first approximation $I = I_g e^{-\tau_0}$ and $I/I_g \simeq 0.15$, we have that $\tau_0 = -\ln(I/I_g) = -\ln(0.15) \doteq 1.9$ and, therefore,

$$T_g = 245\left(1 + \frac{3}{4}1.9\right)^{1/4} \simeq 305°K \qquad (1.3\text{-}10)$$

The temperature obtained in (1.3-10) is somewhat higher than the average temperature on the surface of the earth, but still it describes to a good approximation the greenhouse effect. The small excess we have found in T_g occurs in part because we have neglected the convective transport of heat in the lower atmosphere, which would tend to cool down the surface of the earth (see, e.g., No. 9 of the Bibliography at the end of this chapter). Note that the temperature of the air T_a near the ground is given by (A-30), which yields a value for T_a lower than T_g,

$$T_a = T_e\left(\frac{1}{2} + \frac{3}{4}\tau_0\right)^{1/4} = 245\left(\frac{1}{2} + \frac{3}{4}1.9\right)^{1/4} = 288°K \qquad (1.3\text{-}11)$$

The discontinuity between T_g and T_a is in practice removed through conduction and convection and tends to lower the value of T_g obtained above. Figure 1.3-I describes the balance between the radiation received and the radiation emitted by the earth, including the greenhouse effect.

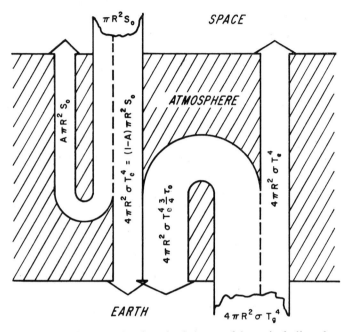

FIGURE 1.3-I   A diagram showing the balance of heat, including the greenhouse effect, in the atmosphere of the earth

It is significant to note that if the earth did not have an atmosphere, or if the terrestrial atmosphere did not have any absorbing molecules such as CO_2, H_2O, and O_3, we would have $\tau_0 = 0$ and $T_g = T_e = 245°K = -28°C = -18.4°F$. This shows the importance of the greenhouse effect, i.e., the trapping of the infrared radiation emitted from the ground by the triatomic molecules of the atmosphere, and emphasizes the critical role of the minor atmospheric constituents. We have already seen the critical role of O_3 in protecting the living organisms on the earth from the harmful ultraviolet radiation of the sun. We should, therefore, be aware of the important role which these minor atmospheric constituents play (like the crucial role that vitamins and hormones play in the human body) and watch carefully their balance, which is not very difficult to upset since they only amount to traces of the terrestrial atmosphere.

Some people have speculated, e.g., that the recent rapid increase of CO_2 in the atmosphere, due to high burning rates of industrial fuels, might enhance the potency of the greenhouse effect and thus increase the average temperature of our planet. This in turn could melt the ice of the polar regions and raise perilously the level of the oceans. Preliminary studies suggest that this is an overexaggerated prediction. The opposite prediction is made by another theory which suggests that the temperature of the earth will drop drastically because increased air pollution (smog, etc.) will increase the terrestrial albedo A which, as seen from (1.3-3) and (1.3-9), will cause a sharp decrease of T_e and therefore of T_g.

Before leaving this subject it should be mentioned that the term "greenhouse effect" is somewhat of a misnomer because, contrary to what is commonly believed, the heating of the greenhouses is not due to the trapping of the ground infrared radiation by the glass roof (like the CO_2 action in the atmosphere) but rather due to the fact that the warm air remains trapped under the glass and does not rise up taking away the heat of the soil. The trapping of the infrared radiation by the glass panes, i.e., the "greenhouse effect" of the glass, amounts to probably less than 20% of the total effect. These results were deduced by comparing greenhouses covered with glass panes and greenhouses covered with rocksalt panes which are transparent to the infrared radiation.

1.4   The Escape of the Atmospheric Gases

When there are no large-scale motions in the atmosphere, we say that the atmosphere is in *hydrostatic equilibrium*, which means that at any given height the pressure of the underlying gases balances the weight of the overlying atmosphere. It is similar to the equilibrium reached when a weight is placed on a bed-spring. Let P be the pressure, T the temperature, ϱ the density, N the number of particles per unit volume, and m the average mass of the particles. All these parameters are interrelated through the equation of state for ideal gases. For unit volume the equation of state is,

$$P = NkT = \frac{\varrho kT}{m} \qquad (1.4\text{-}1)$$

The weight of a pillbox of gas in the atmosphere, as seen from Figure 1.4-I, is $(Adh\,\varrho)\,g$, where g is the acceleration of gravity. This force is counterbalanced by the pressure difference dP below and above the box, so that,

$$(Adh\,\varrho)\,g + A\,dP = 0 \qquad (1.4\text{-}2)$$

Introducing in (1.4-2) the expression for the density ϱ from (1.4-1), we get,

$$\frac{dP}{dh} = -\varrho g = -\frac{mg}{kT}P \qquad (1.4\text{-}3)$$

An approximation which is commonly used is to assume an isothermal atmosphere (T = constant) and neglect the change of the molecular weight and the acceleration of gravity with height. This allows one to introduce as a constant parameter the *scale height H* of the atmosphere,

$$H = \frac{kT}{mg} \tag{1.4-4}$$

and write (1.4-3) in the form,

$$\frac{dP}{P} = -\frac{dh}{H} \tag{1.4-5}$$

which has the solution,

$$P = P_1 e^{-\left(\frac{h-h_1}{H}\right)} \tag{1.4-6}$$

where P_1 is the pressure at the altitude h_1. Under the above stated assumptions, equivalent expressions hold for the density ϱ, and the number of particles per unit volume N. One should realize that this is only an approximation and that H actually varies with height, being approximately 8 km at the ground level and about 100 km at the beginning of the exosphere where $T \simeq 1500°K$, $g \simeq 0.8g_0$, and the mean molecular weight is that of atomic oxygen which predominates around 600 km. Over a certain range of heights, however, the scale height might change very little and, therefore, one can use (1.4-6) successfully in this region using the H that corresponds to this particular range of heights.

The kinetic theory of gases shows that the particle velocities of a gas in thermal equilibrium follow a Maxwellian distribution, which in polar coordinates is given by the expression,

$$Nf(V)\, dV\, d\Omega = N \frac{e^{-(V/V_m)^2}}{(\pi V_m^2)^{3/2}} V^2\, dV \sin\theta\, d\theta\, d\phi \tag{1.4-7}$$

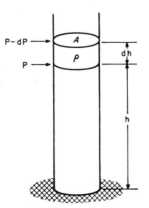

FIGURE 1.4-I A column of atmospheric gas in hydrostatic equilibrium

where V_m is the most probable speed,

$$V_m = \left(\frac{2kT}{m}\right)^{1/2} \tag{1.4-8}$$

When the kinetic energy of a particle exceeds the potential energy of the gravitational field of the earth, this particle can in principle escape to the interplanetary space. The lowest velocity allowing the particle to escape is called the *escape velocity* V_e, and is given by the relation,

$$\frac{1}{2} m V_e^2 = G \frac{mM}{R} = mgR \tag{1.4-9}$$

i.e.,

$$V_e = (2gR)^{1/2} = \left(\frac{2MG}{R}\right)^{1/2} = 11.2 \left(\frac{M}{R}\right)^{1/2} \text{(km/s)} \tag{1.4-10}$$

where M and R are the mass and radius of a given planet expressed in terms of the mass and radius of the earth. Thus, the escape velocity on earth is 11.2 km/s whereas on Mars, with $M = 0.108$ and $R = 0.53$, the escape velocity is ~ 5 km/s. From (1.4-8) and (1.4-10) we see that,

$$\frac{V_e}{V_m} = \frac{(2gR)^{1/2}}{\left(\frac{2kT}{m}\right)^{1/2}} = \left(\frac{R}{kT/mg}\right)^{1/2} = \left(\frac{R}{H}\right)^{1/2} \tag{1.4-11}$$

which for the earth means that $V_e > 8V_m$. Hence, only particles with velocities at the high speed tail of the Maxwellian distribution of (1.4-7) will be able to escape. For $V > 8V_m$ this is only a very small fraction of all the particles.

The requirement $V > V_e$ is not the only condition necessary for the particles to escape. The particles must also be moving upwards and must not be deflected by collisions. Let a number of particles S start moving upwards at an angle θ to the vertical from a height h_1. Of these, only a number S' that has not been stopped or deflected by collisions will reach a higher altitude h_2. The chances for collisions α are proportional to the distance travelled $(h_2 - h_1) \sec \theta$, the ambient density of the particles $N(h)$, and their physical cross-section σ, which is approximately 10^{-15} cm². For an isothermal atmosphere in hydrostatic equilibrium we have,

$$S'(\theta) = S(\theta) \, e^{-\alpha} = S(\theta) \exp \left\{ - \int_{h_1}^{h_2} N(h) \, \sigma \sec \theta \, dh \right\}$$

$$= S(\theta) \exp \left\{ - \int_{h_1}^{h_2} N_1 \, e^{-\left(\frac{h-h_1}{H}\right)} \sigma \sec \theta \, dh \right\}$$

$$= S(\theta) \exp \left\{ - H\sigma N_1 \sec \theta \left[1 - e^{-\left(\frac{h_2-h_1}{H}\right)} \right] \right\} \tag{1.4-12}$$

The probability $p(\theta)$ that a particle will avoid any collision up to its final escape is easily obtained from (1.4-12) by setting $h_2 = \infty$, which yields,

$$p(\theta) = e^{-\alpha} = \frac{S'(\theta)}{S(\theta)} = \exp\left(-N_1 \sigma H \sec \theta\right) \qquad (1.4\text{-}13)$$

From (1.4-13) it follows that the average probability $\bar{p}$, i.e., the probability $p(\theta)$ averaged over all the particles moving in any upward direction is,

$$\bar{p} = \frac{1}{2\pi} \int \int p(\theta)\, d\Omega = \int_0^{\pi/2} \exp\left(-N_1 \sigma H \sec \theta\right) \sin \theta\, d\theta = \int_1^\infty \frac{e^{-\beta\mu}\, d\mu}{\mu^2} \qquad (1.4\text{-}14)$$

where $\beta = N_1 \sigma H$ and $\mu = \sec \theta$. The solution of this integral is,

$$\bar{p} = e^{-\beta} + \beta \left\{ 0.5772157 + \log \beta - \beta + \frac{\beta^2}{2.2!} - \frac{\beta^3}{3.3!} + \frac{\beta^4}{4.4!} \cdots \right\} \qquad (1.4\text{-}15)$$

Numerical evaluation of (1.4-15) shows that when $\beta \simeq 0.25$, $\bar{p} = 50\%$, i.e., half of the particles that have a velocity $V > V_e$ at this height will be able to avoid collisions in their upward motion, and thus will be able to escape the gravitational attraction of the earth. The height h_x at which $\bar{p} = 0.5$ is taken to be the base of the exosphere. The number density N_x at this altitude is,

$$N_x = \frac{\beta}{\sigma H} = \frac{0.25}{10^{-15} 10^7} = 2.5 \times 10^7 \text{ particles/cm}^3 \qquad (1.4\text{-}16)$$

and corresponds to an altitude of $h_x \sim 600$ km.

Practically all of the escaping particles will come from the region near the base of the exosphere. Below this height, because of the increasing density, the escape cone narrows very rapidly and in about one scale height below h_x the much higher collision frequency eliminates the possibility for escape. Above h_x, on the other hand, the probability of escape is excellent and in about 2 scale heights becomes practically 100%, but the particle density of the atmosphere decreases rapidly with height and consequently in these larger heights there are not too many particles to escape.

It should be mentioned that in a more realistic model of the exosphere, where the different gases undergo diffusive differentiation, we have the lighter elements such as hydrogen and helium at the top. These elements, because of their lower m value, have higher thermal velocities for a given temperature ($V^2 = 3kT/m$) and therefore, a larger fraction of these particles has $V > V_e$. Furthermore, at these higher altitudes the gravitational field is weaker and collisions negligible and, therefore, hydrogen and helium have a much higher chance to escape than the heavier nitrogen and oxygen particles which are found mostly in the lower layers of the atmosphere.

To a first approximation, the flux of the escaping particles can be obtained by considering that all the particles of the exosphere are condensed in a thin layer of uniform density N_x having the thickness of one scale height H. Thus the total number of particles in the exosphere is taken to be

$$\eta_x = 4\pi R_x^2 H N_x \tag{1.4-17}$$

It is also assumed that no other particles are left above this layer to deflect the escaping particles. Under these assumptions the total number of particles escaping from the earth in a time interval dt is,

$$d\eta = \left\{ 4\pi R_x^2 \int_0^{2\pi} \int_0^{\pi/2} \int_{V_e}^{\infty} (V \cos \theta)\, N_x f(V)\, dV\, d\Omega \right\} dt \tag{1.4-18}$$

where $N_x f(V)\, dV\, d\Omega$ is the Maxwellian distribution of the velocities given by (1.4-7). Integration of (1.4-18), taking into consideration that $(V_e/V_m)^2 \gg 1$, yields,

$$d\eta \simeq \left\{ 4\pi R_x^2 \frac{N_x}{2\pi^{1/2}} \frac{V_e^2}{V_m} e^{-(V_e/V_m)^2} \right\} dt \tag{1.4-19}$$

and therefore, dividing (1.4-19) by (1.4-17) we find that the fraction of the exospheric particles escaping in time dt is,

$$\frac{d\eta}{\eta_x} = - \left\{ \frac{V_e^2}{2\pi^{1/2} V_m H} e^{-(V_e/V_m)^2} \right\} dt \tag{1.4-20}$$

If the number of particles in the exosphere η_x is now only a fraction X of the total number of particles in the entire atmosphere η, i.e.,

$$\eta_x = X\eta \tag{1.4-21}$$

we can introduce (1.4-21) in (1.4-20) to get,

$$\frac{d\eta}{\eta} = \frac{-V_e^2 X}{2\pi^{1/2} H V_m \exp (V_e/V_m)^2}\, dt \tag{1.4-22}$$

Finally, by integrating (1.4-22) we obtain,

$$\eta = \eta_0 e^{-(t/t_0)} \tag{1.4-23}$$

where η_0 is the number of particles originally present in the entire atmosphere and t_0 is the *escape time*, i.e., the time required to reduce the particle density to $1/e$ of η_0. From (1.4-22) and (1.4-23) it follows that,

$$t_0 = \frac{V_m}{gX_0} \frac{\exp (R_x/H)}{(R_x/H)} \tag{1.4-24}$$

where in (1.4-24) we have incorporated the factor $\pi^{1/2}$ in the parameter X_0 and we have used (1.4-4), (1.4-8), and (1.4-11) to simplify the expression for t_0.

Typical values of X_0 are of the order of 10^{-6}, but differ significantly for the different atmospheric constituents and are of course different for the different planets of the solar system. It is interesting to note that in the case of the moon, X_0 is of the order of unity because any atmosphere that the moon might have had should be essentially classified as exosphere. Table 1.4-I gives the escape time t_0 in years for H, He, O, N_2, and A, for the moon, the earth, Venus, Mars, and Jupiter.

TABLE 1.4-I Escape time in years

	Moon	Earth	Mars	Venus	Jupiter
H	10^{-3}	$10^{3.5}$	10^2	10^3	10^{500}
He	10^{-2}	10^8	10^3	10^5	10^{2000}
O	10^1	10^{35}	10^9	10^{25}	—
N_2	10^2	10^{60}	10^{17}	10^{40}	—
Ar	10^9	10^{80}	10^{25}	$10^{80\circ}$	—

1.5 The Atmospheres of the Planets

The exponents in the figures of Table 1.4-I have an uncertainty of about 10% for the earth and about 25% or more for the other planets. The reason is that the most important factor in the expression for t_0, namely $\exp(R/H)$, has an exponential dependence on the temperature because $H = kT/mg$. As a result, a small uncertainty in the temperature becomes a large uncertainty in t_0. The average exospheric temperature of the earth is $\sim 1500°K$, which is subject to a small averaging uncertainty due to the already discussed latitudinal and periodic variations. The average exospheric temperatures of the other planets are known only approximately and contain probably much larger errors. In spite of these uncertainties, however, there are certain obvious results that can be deduced from Table 1.4-I. Thus the moon, with an escape velocity of 2.5 km/s, has lost all the atmosphere that it might have had, while Jupiter, with its low exospheric temperature and high escape velocity, has retained practically all of its original atmosphere.

All the terrestrial planets have lost their hydrogen, and the small amounts of hydrogen present (as, e.g., the protonosphere of the earth) are of secondary origin coming most probably from the solar wind. Mars and Venus have also lost their helium, while the earth is near the border line. Oxygen

2*

on Mars is also a borderline case, while earth and Venus have been capable of retaining whatever free oxygen was present. The Jovian planets (Jupiter, Saturn, Uranus, and Neptune), with their large escape velocities and low temperatures, have retained their original atmospheres which, as seen from spectral studies, are rich in H_2, CH_4, and NH_3.

Methane and ammonia were also present in the primordial atmosphere of the earth and played a very essential role, at least according to one of the more widely held theories, in the formation of the first organic compounds. CH_4 and NH_3, however, are easily decomposed by the ultraviolet radiation in the upper layers of the atmosphere, and the hydrogen can escape in the interplanetary space. This probably happened before CH_4 and NH_3 could be oxidized by the free oxygen which appeared later when the atmosphere of the earth changed from reducing to oxidizing.

The same would have been the fate of the atmospheric water vapor except for some very fortunate conditions which occur in the terrestrial atmosphere and have allowed our planet to preserve its water supply. First of all, H_2O boils (and condenses) at a relatively high temperature and, therefore, water vapor changes to water droplets without the need of very low temperatures or very high pressures. In the case of the earth, the temperature of the troposphere falls rapidly with altitude and before the tropopause it reaches a low enough value to produce the condensation of water vapor to water droplets which return to the ground.

This is called the *cold trap* of the atmosphere, which does not allow H_2O to reach heights above 10 km. On the other hand, the ultraviolet radiation of the sun, which could decompose the water vapor, is stopped at higher altitudes by the molecular oxygen and the ozone layer of the atmosphere and cannot penetrate to heights below 10 km to dissociate the water vapor. As a result of all these fortunate circumstances, the water supply of the earth with all its important consequences has been preserved. In the case of Venus, an ineffective cold trap (higher temperatures) and insufficient shielding from the solar UV rays (lack of oxygen and ozone) have probably combined to permit the depletion of practically all of its water supply. The lack of water, as we have seen, has allowed Venus to accumulate its large atmosphere of CO_2.

Following is a brief summary of the atmosphere of the other planets:

Mercury Until a few years ago it was thought that Mercury always keeps the same side locked on the sun, exactly like the moon does with the earth. Consequently it was believed that the temperature at the subsolar point was about 620°K (above the melting point of lead), while the dark side of Mercury remained near absolute zero at a temperature of about 30°K.

Recent radar observations, however, have shown that Mercury rotates with a period of about 59 days, which is 2/3 of its 88-day year. As a result, it now appears that the subsurface temperature, i.e., the temperature below the very top ground layer, does not change that drastically from the sunlit to the dark side of the planet. This view is supported also by recent microwave observations of Mercury. If Mercury has any atmosphere at all, it will be a very thin one consisting probably of Ar^{40} from the radioactive decay of K^{40}, krypton, xenon, and possibly some CO_2.

Venus   The Cytherian planet is always covered with clouds which prevent us from seeing the surface of the planet and computing, from the rotation of different landmarks, the rotational period of Venus. Only recently radar observations have shown that Venus rotates in a retrograde fashion, i.e., in the opposite sense than practically all the other members of the solar system, and that it completes a rotation around its axis in 243 ± 0.5 days. As seen from Figure 1.4-I, this is probably the 243.16 day resonance between Venus and earth which requires Venus always to show the same face to the earth when the two planets meet at *inferior conjunction*.

Inferior conjunction of earth and Venus occurs when the earth, Venus, and the sun are all aligned in this order. The time interval between successive inferior conjunctions is called the *synodic period* S and is related to the *sidereal year* of the earth Y_E and of Venus Y_V by the expression,

$$\frac{1}{S} = \frac{1}{Y_V} - \frac{1}{Y_E} \qquad (1.5\text{-}1)$$

With $Y_V = 244.7$ days and $Y_E = 365.25$ days, we find $S = 584$ days. If the retrograde rotational period of Venus is $R_V = 243.16$ days, then the Venusian day (e.g., from sunrise to sunrise or from noon to noon at a given point on Venus) is given by the expression,

$$\frac{1}{D_V} = \frac{1}{Y_V} + \frac{1}{R_V} \qquad (1.5\text{-}2)$$

which gives $D_V = 116.8$ days. In (1.5-2) we have used the plus sign for the $1/R_V$ because the rotation of Venus is retrograde. This constitutes a resonance because the synodic period S is exactly equal to 5 Venusian days D_V ($584 = 5 \times 116.8$). As a result, at inferior conjunction it is always the same geographical points of Venus that have noon on the side facing the sun and midnight on the dark side facing the earth. Thus, as seen from Figure 1.5-I, the earth always sees the same face of Venus at inferior conjunction.

The ground temperature of Venus is approximately 700°K. This was originally deduced from radio astronomical observations in the cm-range from the earth and was later confirmed by the flyby missions of the Mariner spaceships. The most exciting results, however, were the in situ measurements obtained by several of the Venera spacecrafts of the Soviet Union which parachuted capsules of instruments into the atmosphere of Venus.

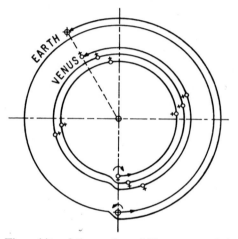

FIGURE 1.5-I   The orbits of the earth and Venus around the sun. Note that Venus rotates about its axis in retrograde motion

The results of these observations, coupled with terrestrial observations in the radio, visible, and infrared portions of the spectrum, have shown that the atmosphere of Venus consists almost exclusively of CO_2 ($\sim 95\%$). The nitrogen content is definitely less than 7% since the instruments of the Venera 4 capsule, with a 7% threshold for nitrogen, gave negative results. It is probably only 3–4% which, with a ground pressure of 80–100 Atm., gives a total amount of nitrogen three to four times the amount found in the terrestrial atmosphere, but approximately equal to the total amount of nitrogen found in both the atmosphere and the crust of the earth. Free oxygen seems to be practically absent from the atmosphere of Venus while the water vapor content is approximately 1%, i.e., about 1 Atmosphere.

It is interesting to note that if all the water and ice on the surface of the earth was evaporated it would produce a ground pressure of at least 100 Atmospheres. It is then clear that Venus has lost practically all of its water due to its higher temperature (Jastrow, 1968). Without water, as we have seen, the CO_2 cannot be withdrawn from the atmosphere and as a result all the CO_2 which was outgassed over the past history of Venus has accumulated to produce its present monstrous CO_2 atmosphere.

The very high temperature on the surface of Venus is most probably due to the strong greenhouse effect of the CO_2. The small amount of water vapor present is probably contributing also by filling in the windows of the CO_2 absorption spectrum. Without the greenhouse effect, the closer proximity of Venus to the sun would only make its temperature 15% higher than the temperature of the earth. Friction heating caused by the very strong winds (several hundred miles per hour) which appear to blow continuously on the desert-like surface of Venus might be also a contributing factor toward the very high temperatures observed.

At an altitude of approximately 25 km, the Venera capsules found that the temperature ($\sim 40°C$) and pressure (~ 1 Atm.) of Venus resemble the conditions prevailing on the surface of the earth. As a result, it is not totally inconceivable that balloonlike organisms, filled probably with hydrogen, can still find a hospitable environment on Venus. Above this altitude there is a thick layer of clouds which cause the high (0.75) albedo of Venus. The temperature at the bottom of the clouds is $\sim 300°K$ and at the top, $\sim 240°K$. This last temperature is obtained from infrared observations in the 8–12 μ window of the terrestrial atmosphere.

Both the American and the Russian spacecrafts have found that Venus has an ionosphere with a maximum of about 10^5 el/cm^3, but lacks a magnetic field and, therefore, does not have a magnetosphere.

Mars   The atmospheric pressure on the surface of Mars has been deduced from radio occultation experiments that were conducted in the fly-by missions to Mars of Mariner 4 in 1965 and of Mariners 6 and 7 in 1969 (Kliore *et al.*, 1969). It was found to be approximately 5 millibars, i.e., about 0.5% of the atmospheric pressure on the surface of the earth. A probable explanation for its very thin atmosphere is that Mars, being considerably smaller, has a very small, if any, molten core and, therefore, the outgassing activity on Mars has been very limited. The ground temperature on Mars varies from a comfortable 300°K at the subsolar point to a deepfreezing 200°K or less in the polar regions. The atmosphere of Mars consists predominantly of CO_2, which has been clearly identified.

It must also contain some N_2, which is difficult to detect spectroscopically because it lacks absorption bands in the portion of the spectrum that can be analyzed with ground telescopes. Argon and neon are probably also present in considerable amounts. Very small amounts of H_2O have already been observed and the spectrometers on recent Mariner missions have also detected traces of oxygen, hydrogen, and carbon monoxide. The very low abundance of oxygen implies the lack of ozone and, therefore, the Martian atmosphere will not show the temperature increase which occurs in the terrestrial stratosphere due to the absorption of the ultraviolet radiation by the ozonosphere. The polar caps of Mars are now believed to be made of dry ice, i.e., solid CO_2, plus possibly some solid H_2O.

Mars does not have a magnetic field either, but it has an ionosphere with a maximum of $1-2 \times 10^5$ el/cm^3 at a height of approximately 135 km.

The Asteroids   These are either the fragments of a demolished planet or chunks of matter that never managed to stick together to form a planet. Most of them circle the sun in a belt between Mars and Jupiter. The largest of them are Ceres and Pallas, but even these are too small (~ 400 miles in diameter) to hold any atmosphere. It would be of great interest, however, in the study of the history of the solar system to analyze the gases trapped in the rocks of the asteroids.

Jupiter   Infrared and radio observations yield temperatures in the range of 130–150°K and refer to the top of the Jovian cloud layer, which most probably consists of NH_3 crystals. These temperatures are somewhat higher than the effective temperature of ~ 100°K which is obtained from the value of the solar constant at the orbit of Jupiter and the Jovian albedo which is ~ 0.5 in the optical portion of the spectrum. This difference suggests that Jupiter might possess an additional internal heat source (Rasool, 1968). The scale height of the Jovian atmosphere was computed from an occultation of the star σ-Arietis by Jupiter and was found to be ~ 8 km. From the known expression of $H = kT/mg$, the value $H = 8$ km corresponds to a mean molecular weight of approximately 4. Jupiter, with its low temperature and high gravity, has retained all of its original atmosphere with the large abundance of hydrogen and helium which are responsible for the very low mean molecular weight. NH_3 and CH_4 have also been detected on Jupiter in substantial amounts.

What goes on below the clouds of Jupiter is very difficult to infer. Some authors have suggested that Jupiter is completely covered with ice, others with liquid ammonia, while some have suggested a dense gas sphere at temperatures as high as 1000°K and pressures as high as several thousand atmospheres. Some very interesting features of the Jovian atmosphere are its band structure and the famous Red Spot, neither of which is well understood. Jupiter possesses a strong magnetic field and radiation belts (Van Allen belts) which emit intense synchrotron radiation. The value of the Jovian magnetic field near the surface of the planet is approximately 10 times higher than the ground balue of the terrestrial magnetic field.

Saturn, Uranus, and *Neptune*   All three are massive (giant) planets very similar to Jupiter. Their respective temperatures are near or below 100°K and decline with the planet's distance from the sun. Their atmospheres consist basically of H_2 and He and undoubtably also contain CH_4 and NH_3. They all have albedos similar to that of Jupiter and most likely they are also covered with clouds. This is by no means certain, however, for Uranus and Neptune, where the observed albedo might be due to Rayleigh scattering in a dense atmosphere rather than due to reflecting clouds.

Pluto This small, far-away planet was discovered only in 1930 from the small perturbations it causes to the orbits of Uranus and Neptune. Pluto at one time might have been a satellite of Neptune. Because of the large distance and small size (similar to the terrestrial planets) of Pluto, we know very little about its atmosphere. The temperature of Pluto must be very low ($\sim 50°K$) because of its great distance (~ 39.5 Astronomical Units) from the sun.

TABLE 1.5-I

Name	Symbol	Distance in A.U.	Radius in R_0	Mass in M_0	Gravity in m/s²	Esc. Vel. in km/s	Albedo
Mercury	☿	0.387	0.38	0.054	3.6	4.2	0.06
Venus	♀	0.723	0.96	0.815	8.7	10.3	0.75
Earth	⊕	1.000	1.00	1.000	9.8	11.2	0.4
Mars	♂	1.524	0.53	0.108	3.8	5.0	0.15
Ceres (Asteroid)		2.767	0.055	0.0001	0.3	0.5	0.07
Jupiter	♃	5.203	11.19	317.8	26.0	61.0	0.5
Saturn	♄	9.540	9.47	95.2	11.2	37.0	0.5
Uranus	♅	19.180	3.73	14.5	9.4	22.0	0.5
Neptune	♆	30.070	3.49	17.2	15.0	25.0	0.5
Pluto	P	39.440	~ 0.4	~ 0.2	~ 12.3	~ 7.6	~ 0.4

1 A.U. $= 1.5 \times 10^{13}$ cm
1 $R_\oplus$ $= 6.38 \times 10^8$ cm
1 $M_\oplus$ $= 5.48 \times 10^{27}$ gr

1.6 Bibliography

A. Books for Further Studies

1. *Structure and Evolution of the Stars*, M. Schwarzschild, Princeton University Press, Princeton, N. J., 1958.
2. *Astrophysics and Space Science*, A. J. McMahon, Prentice Hall, Englewood Cliffs, N. J., 1965.
3. *Solar System Astrophysics*, J. C. Brandt and P. W. Hodge, McGraw Hill, New York, 1964.
4. *Space Physics*, R. S. White, Gordon and Breach, New York, 1970.

5. *The Origin of the Solar System*, ed. by R. Jastrow and A. G. W. Cameron, Academic Press, New York, 1963.
6. *Intelligent Life in the Universe*, I. S. Shklovskii and C. Sagan, Holden-Day, Inc., San Francisco, Calif., 1966.
7. *The Atmospheres of the Earth and the Planets*, ed. by G. P. Kuiper, The University of Chicago Press, Chicago, Ill., 1952.
8. *Atmospheric radiation*, R. M. Goody, Clarendon Press, Oxford, Great Britain, 1964.
9. *Space Exploration and the Solar System*, Course XXIV, ed. by B. Rossi, Academic Press, New York, 1964 (Planetary Atmospheres by R. Jastrow).
10. *An Introduction to Atmospheric Physics*, R. G. Fleagle and J. A. Businger, Academic Press, New York, 1963.
11. *The Upper Atmosphere: Meteorology and Physics*, R. A. Craig, Academic Press, New York, 1965.
12. *Structure of the Stratosphere and Mesosphere*, W. L. Webb, Academic Press, New York, 1966.
13. *Physics of the Upper Atmosphere*, ed. by J. A. Ratcliffe, Academic Press, New York, 1960.
14. *Introduction to Space Science*, ed. by W. N. Hess and G. D. Mead, Gordon and Breach, New York, 1968.
15. *Space Physics*, ed. by D. P. LeGalley and A. Rosen, John Wiley and Sons, New York, 1964.
16. *Research in Geophysics*, ed. by H. Odishaw, M. I. T. Press, Cambridge, Mass., 1964.
17. *Geophysics the Earth's Environment*, ed. by G. DeWitt, J. Hieblot and A. Lebeau, Gordon and Breach, New York, 1963.
18. *Handbook of Geophysics and Space Environments*, ed. by S. L. Valley, Air Force Cambridge Res. Lab., Bedford, Mass., 1965.
19. *Satellite Environment Handbook*, ed. by F. S. Johnson, Stanford University Press, Stanford, Calif., 1965.
20. *U. S. Standard Atmosphere* 1962, U. S. Government Printing Office, Washington, D. C., 1962.
21. *U. S. Standard Atmosphere* 1966, U. S. Government Printing Office, Washington, D. C., 1966.

B. Articles in Scientific Journals

Cloud, P. E., Atmospheric and hydrospheric evolution of the primitive Earth, *Science*, **160**, 729, 1968.
Berkner, L. V. and L. C. Marshall, The history of growth of oxygen in the Earth's atmosphere, *Progress in Radio Science* 1960–1963, vol. III, p. 174, ed. by G. M. Brown, Elsevier Publ. Co., New York, 1965.
Evans, J. V., Radar signatures of the planets, *Annals of N. Y. Academy of Sciences*, **140**, 196, 1966.
Jacchia, L. G., Recent advances in upper atmospheric structure, *Space Research* **X**, 367, 1970.
Jastrow, R., The planet Venus, *Science*, **160**, 1403, 1968.
Kliore, A., G. Fjeldbo, B. Seidel, and S. Rasool, Mariner 6 and 7: Radio occultation measurements of the atmosphere of Mars, *Science*, **166**, 1393, 1969.
Ponnamperuma, C. and N. W. Cabel, Current status of chemical studies on the origin of life, *Space Life Sciences*, **1**, 64, 1968.

Rasool, S. I., Jupiter Rosetta Stone of the solar system, *Astronautics and Aeronatuics*, October 1968.

Revelle, R. *et al.*, Atmospheric carbon dioxide, *Report of the Environment Pollution Panel*, President's Science Advisory Committee, The White House, pp. 111–133, 1965.

Sagan, C., Gray and non-gray planetary atmospheres. Structure, convective instability and greenhouse effect. *Icarus*, **10**, 290, 1969.

CHAPTER 2

THE IONOSPHERE

2.1 Introduction

The ionosphere is the partially ionized region of the terrestrial atmosphere extending from approximately 70 km on up. The existence of the ionosphere, as an electrically conducting region of the atmosphere, was first suggested by the Scottish meteorologist Belfour Stewart in 1883. Marconi's successful experiments in 1901 of wireless communication across the Atlantic prompted Heaviside and Kennelly to postulate independently the existence of an ionized layer in the atmosphere. This electrically conducting layer was originally called the Heaviside layer and later the E-layer because of its many free electrons. The E-layer, as seen in Figure 2.1-I, acts as a reflector and makes it possible for radio signals to bridge large distances over the spherical earth. The E-layer is at an altitude of approximately 110 km.

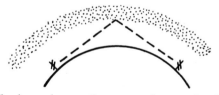

FIGURE 2.1-I   The ionosphere acting as a reflector of radio waves making possible radio telecommunication over the horizon

The first attempts to study the structure of the ionosphere with radio signals bounced back from the ionosphere were made in 1925 by Appleton and Barnett in England. Similar ionospheric sounding experiments were performed also in America in 1928 by Brest and Tuve. An *ionospheric sounder* consists basically of a radio transmitter and a radio receiver connected in a way which allows them to measure the time interval between the transmission and the return of the radio pulse. By multiplying one half of this time interval, which is of the order of a millisecond, with the speed of light, we obtain the height of the reflection layer. These early sounding experiments showed that above the *E-layer* there is another layer with even higher electron density. This layer, following the alphabetical order, was named the *F-layer*. Later it was also found that during the day the *F*-layer separates into two layers which were named the *F1-layer* and the *F2-layer*. Finally a weak layer was discovered, this one below the *E*-layer, and following the tradition it was named the *D*-layer. It should be noted that these layers represent only small enhancements in the electron density profile of the ionosphere and they do not constitute separate layers. For this reason in the recent literature the term "layer" is frequently replaced with the term "region" (*D-region, E-region, F2-region*).

The ionization of the atmosphere is produced primarily by the sun's ultraviolet and x-ray radiation. The rate q at which ion-electron pairs are produced per unit volume is proportional to the intensity of the ionizing radiation I and the number density N_n of the neutral atmosphere, i.e.,

$$q \propto I N_n \tag{2.1-1}$$

As seen from the schematic diagram of Figure 2.1-II, at high altitudes q is very small because N_n is very small. As the ionizing radiation penetrates deeper into the more dense layers of the atmosphere, q reaches a maximum q_m at a height h_m where I and N_n reach the best possible combination. Below this altitude, the intensity of the ionizing radiation drops rapidly because the energy is spent for the ionization of the atmosphere. As I decreases, q also decreases (2.1-1) and finally vanishes near 70 km.

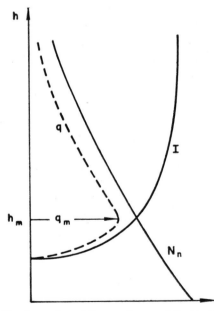

FIGURE 2.1-II The formation of the maximum of the ionosphere at the height of the most favorable balance between the ionizing radiation I and the particle density of the atmosphere N_n

Chapman, in 1931, produced a very neat theoretical treatment of the problem. In his simplified model, Chapman assumed an isothermal, horizontally stratified atmosphere, composed of a single gas which is being ionized by a monochromatic radiation from the sun. It is obvious that this model is an oversimplification of the actual conditions. Simplifications of this kind, however, are always done in order to formulate a mathematically workable theory. It is a real test to the ingenuity of the investigator to build a model simple enough to have a tractable mathematical solution but at the same time realistic enough to yield physically meaningful results. The Chapman layer theory (1931 a, b) is a very good example of an ingenious mathematical formulation of a very complicated physical problem.

2.2   The Chapman Layer Theory

In section 1.4 we have seen that the density ϱ of an isothermal atmosphere in hydrostatic equilibrium varies exponentially with height, i.e.,

$$\varrho(h) = \varrho_g \, e^{-h/H} \tag{2.2-1}$$

where ϱ_g is the density at ground level and H is the constant scale height. Let us first compute the absorption sustained by a beam of ionizing radiation at a height h. Let the beam have unit cross-section and let χ be the angle the beam makes with the vertical (zenith angle). The energy of the beam expended to ionize neutral particles between h and $h + dh$ (Figure 2.2-I) will be proportional to the intensity of the beam at this height $I(h)$, and to the amount of ionizable matter $\varrho(h)\,dh\,\sec\chi$ encountered in this layer. If σ is the constant of proportionality in this relation, then the amount

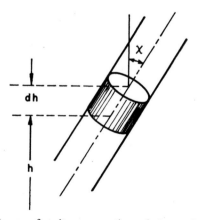

FIGURE 2.2-I   A beam of unit cross-section of the sun's ionizing radiation incident at a zenith angle χ at a height h of the earth's atmosphere

of radiation absorbed in this layer will be,

$$dI = \sigma I(h)\, \varrho(h) \sec \chi\, dh = \sigma I(h)\, \varrho_g e^{-h/H} \sec \chi\, dh \qquad (2.2\text{-}2)$$

Integrating (2.2-2) from the height h to infinity (actually to the top of the atmosphere) we get,

$$I(h) = I_\infty \exp\left(-\sigma \varrho_g \sec \chi\, H e^{-h/H}\right) \qquad (2.2\text{-}3)$$

where I_∞ is the intensity of the ionizing radiation at the top of the atmosphere. The rate q at which ion-electron pairs are produced per unit volume is equal to the energy absorbed per unit volume because according to the simplifying assumptions we made (monochromatic radiation and a single kind of ionizable particles) each photon lost produces an electron-ion pair. Thus, since dI was absorbed in a cylinder of unit cross-section and of length $\sec \chi\, dh$ we have,

$$q = \frac{dI}{\sec \chi\, dh} = \sigma I(h)\, \varrho_g e^{-h/H}$$

$$= \sigma \varrho_g I_\infty \exp\left\{ -\frac{h}{H} - \sigma \varrho_g \sec \chi\, H e^{-h/H} \right\} \qquad (2.2\text{-}4)$$

We can now find the height of maximum electron production h_m, by equating to zero the derivative of (2.2-4) with respect to h. The result is,

$$h_m = H \ln\left(\sigma \varrho_g \sec \chi\, H\right) \qquad (2.2\text{-}5)$$

Introducing (2.2-5) in (2.2-4) we get,

$$q = \sigma \varrho_g I_\infty \exp\left\{ -\frac{h}{H} - e^{h_m/H}\, e^{-h/H} \right\}$$

$$= \frac{I_\infty}{H \sec \chi} \exp\left\{ -\left(\frac{h - h_m}{H}\right) - e^{-(h-h_m)/H} \right\} \qquad (2.2\text{-}6)$$

Since the maximum production of electrons q_m occurs at the altitude $h = h_m$, we find from (2.2-6) that,

$$q_m = \frac{I_\infty e^{-1}}{H \sec \chi} \qquad (2.2\text{-}7)$$

In order to simplify the appearance of (2.2-6) we often use the dimensionless parameter,

$$z = \left(\frac{h - h_m}{H}\right) \qquad (2.2\text{-}8)$$

and by introducing (2.2-7) and (2.2-8) in (2.2-6) we obtain the basic formula of the Chapman layer theory,

$$q = q_m \exp\left(1 - z - e^{-z}\right) \qquad (2.2\text{-}9)$$

The change of the electron content per unit volume per unit time, i.e. dN/dt, is equal to the production rate q minus the loss rate L. Electrons are lost from a given volume by recombination, by attachment, and by moving out. The rate of recombination is proportional to the number of electrons and the number of positive ions present, i.e. $L_R \propto N_e N_i$. The constant of proportionality of this relation is usually symbolized with the letter α which is called the *recombination coefficient*. Thus we have,

$$L_R = \alpha N^2 \qquad (2.2\text{-}10)$$

because in the ionosphere which is electrically neutral we have essentially $N_e = N_i = N$, where N stands for either N_e or N_i.

Some electrons are also lost by attaching themselves to neutral atoms, which are in great abundance and therefore do not bottleneck the reaction. As a result, the rate of loss by attachment L_A is proportional only to the number of electrons present per unit volume. The proportionality constant is called the *attachment coefficient* and is usually symbolized with the letter β. Thus we have,

$$L_A = \beta N \qquad (2.2\text{-}11)$$

Actually, we will soon see that this loss mechanism is not true attachment but rather an attachment-like loss process in the sense that it is described mathematically as if it were attachment. Finally the loss rate due to the motion of the electrons is,

$$L_M = \nabla \cdot (NV) \qquad (2.2\text{-}12)$$

where V is the drift velocity of the electrons. Naturally, this term could also represent a gain in certain cases. Summarizing all the above results, we can write the continuity equation of the electron density N in the following form,

$$\frac{dN}{dt} = q - L \qquad (2.2\text{-}13)$$

For quasi-equilibrium conditions, the electron density N is nearly constant (changes very slowly) so that dN/dt tends to zero. We can also to a first approximation neglect the attachment loss and the drift motion of the electrons so that (2.2-13) will become $q = L_R$ which yields,

$$N = \left(\frac{q}{\alpha}\right)^{1/2} = \left(\frac{q_m}{\alpha}\right)^{1/2} \exp\frac{1}{2}(1 - z - e^{-z}) \qquad (2.2\text{-}14)$$

At $h = h_m$, i.e. at $z = 0$, we have the maximum production of electrons and therefore the maximum electron density N_m. Setting $z = 0$ in (2.2-14) we obtain,

$$N_m = \left(\frac{q_m}{\alpha}\right)^{1/2} = \left\{\frac{I_\infty \, e^{-1} \cos \chi}{\alpha H}\right\}^{1/2} \qquad (2.2\text{-}15)$$

and by introducing (2.2-15) in (2.2-14), we obtain the famous expression for the type-α Chapman electron density profile,

$$N = N_m \exp \frac{1}{2}(1 - z - e^{-z}) \qquad (2.2\text{-}16)$$

Similar expressions can be obtained for the region where attachment is the most important loss mechanism, and the resulting distribution is called the type-β Chapman profile.

As seen from (2.2-5) and (2.2-15), a change of the zenith angle χ produces a change in N_m and h_m. This amounts to changing the height of N_m and multiplying the entire profile by a constant factor, namely $(\cos \chi)^{1/2}$. These changes, however, do not alter the basic form of the electron density profile which, in a logarithmic plot, retains the same shape for any zenith angle χ. The curvature K of the Chapman profile at $N = N_m$ is,

$$K = \frac{N'(h)}{\{1 + N'(h)^2\}^{3/2}}\bigg|_{N=N_m} = \frac{N_m}{2H^2} \qquad (2.2\text{-}17)$$

where $N'(h)$ is the derivative of $N(h)$, given by (2.2-16), with respect to h at $N = N_m$, i.e., at $z = 0$.

A parabolic electron density profile of half thickness $2H$ which is given by the expression,

$$N = N_m\left\{1 - \left(\frac{h - h_m}{2H}\right)^2\right\} = N_m\left\{1 - \left(\frac{z}{2}\right)^2\right\} \qquad (2.2\text{-}18)$$

has the same curvature $(K = N_m/2H^2)$ at $N = N_m$ and because of its mathematical simplicity is often used as a substitute for the more complicated Chapman profile. Of course, one can also readily see that the Chapman profile reduces to a parabolic profile for small values of z, i.e. near N_m,

$$\exp \frac{1}{2}(1 - z - e^{-z}) \rightarrow \exp \frac{1}{2}\left\{1 - z - \left(1 - z + \frac{z^2}{2}\right)\right\}$$

$$= \exp\left(-\frac{z^2}{4}\right) \rightarrow \left\{1 - \left(\frac{z}{2}\right)^2\right\} \qquad (2.2\text{-}19)$$

When the zenith angle χ is equal to zero, the maximum electron density, which for $\chi = 0$ we will denote by N_0, occurs at a height h_0 which, as seen from (2.2-5), is given by the expression,

$$h_0 = H \ln (\sigma \varrho_g H) \qquad (2.2\text{-}20)$$

From (2.2-5) and (2.2-20) follows that,

$$(h_m - h_0) = H \ln (\sec \chi) \qquad (2.2\text{-}21)$$

which shows clearly that the maximum electron density occurs at the lowest possible height near noon when the zenith angle χ has its smallest value. Using the dimensionless parameter z_0,

$$z_0 = \left(\frac{h - h_0}{H}\right) \tag{2.2-22}$$

we can obtain the following equations which are equivalent to the ones we have already derived.

$$q = q_0 \exp\left(1 - z_0 - \sec \chi \, e^{-z_0}\right) \tag{2.2-23}$$

$$q_0 = \left(\frac{I_\infty \, e^{-1}}{H}\right)^{1/2} \tag{2.2-24}$$

$$N = N_0 \exp\frac{1}{2}\left(1 - z_0 - \sec \chi \, e^{-z_0}\right) \tag{2.2-25}$$

$$N_0 = \left(\frac{q_0}{\alpha}\right)^{1/2} = \left(\frac{I_\infty \, e^{-1}}{\alpha H}\right)^{1/2} = \frac{N_m}{\cos^{1/2} \chi} \tag{2.2-26}$$

Figure 2.2-II is a plot of the ratio N/N_0 versus z_0, for $\chi = 0°$, $30°$, $60°$, and $80°$. The corresponding parabolic profile for $\chi = 0$ is also shown in Figure 2.2-II with a dashed line. Equations (2.2-25) and (2.2-26) refer to the type-α Chapman profile. The corresponding expressions for the type-β Chapman electron density profile are,

$$N = N_0 \exp\left(1 - z_0 - \sec \chi \, e^{-z_0}\right) \tag{2.2-27}$$

$$N_0 = \left(\frac{q_0}{\beta}\right) = \left(\frac{I_\infty \, e^{-1}}{\beta H}\right) = \frac{N_m}{\cos \chi} \tag{2.2-28}$$

In the following section we will see that the critical frequency of the ionosphere, i.e. the highest frequency which is still reflected by the ionosphere, is proportional to the square root of the electron density, i.e. $f_m \propto N_m^{1/2}$. Thus from the last relation of (2.2-26) we find that,

$$I_\infty \propto \left(\frac{N_m}{\cos^{1/2} \chi}\right)^2 \propto \left(\frac{f_m^4}{\cos \chi}\right) \tag{2.2-29}$$

This means that the expression $(f_m^4/\cos \chi)$, which is called the *character figure* of the ionosphere, is proportional to the intensity I_∞ of the ionizing radiation. The intensity I_∞ is in general proportional to the *sunspot number*, which is a semi-empirical index of the solar activity. Plots of the E-layer character figure $(f_0E)^4/\cos \chi$ vs. the sunspot number $(\bar{R})$ produce during the 11 year cycle of the solar activity a straight line which is given by the relation,

$$\frac{(f_0 E)^4}{\cos \chi} = 120 + \bar{R} \tag{2.2-30}$$

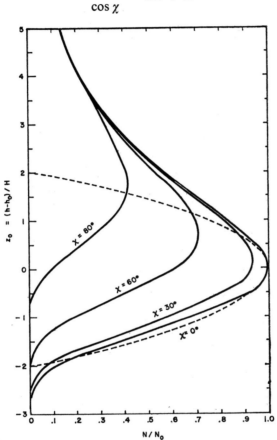

FIGURE 2.2-II   The change of N/N_0 with $z_0 = (h - h_0)/H$ at different values of the zenith angle χ for the type-α Chapman profile. The parabolic profile for $\chi = 0$ is shown in dashed line

Similar plots for the critical frequency of the $F2$-layer ($f_0 F2$) give results which agree more with the predictions for a type-β Chapman profile (2.2-28), i.e.,

$$\left(\frac{N_m}{\cos \chi}\right) \propto \left(\frac{f_m^2}{\cos \chi}\right) \propto I_\infty \tag{2.2-31}$$

The agreement, however, is not as good as it was for the E-layer with a type-α profile because the $F2$-region is much more complex and a simple theory, like the Chapman layer theory, cannot provide a very adequate description.

3*

2.3   The Plasma Frequency

Practically all of our knowledge about the ionosphere has come through radio sounding. Only in the late fifties and early sixties some measurements of local electron densities were made in the upper ionosphere using rockets and satellites equipped with Langmuir probes, Faraday Cups, etc., but even these methods have now been abandoned in favor of the more efficient *top-side sounder* satellites which again use radio waves to probe the top side of the ionosphere.

Let us consider an ionized layer with a uniform electron density N and radio waves of frequency f incident normally (at right angles) upon the layer. If the frequency is above a limiting frequency f_N the waves will pass through the layer, whereas if $f < f_N$, the waves will be reflected back. This critical frequency is called the *plasma frequency* f_N and is proportional to the square root of the electron density N of the layer.

Plasma is the name given to a mixture of electrons, ions, and neutral particles. When an electromagnetic wave such as the radio wave enters into a plasma, its electric field tends to set the charged particles in motion. The ions, which are about 10^4 times heavier than the electrons, respond very little to the weak field of the wave and can be considered as stationary. The light electrons, on the other hand, react readily to the $-eE$ force acting on them ($-e$ is the negative charge of the electron). Let N_i and N_e be the initial (unperturbed) number densities of the ions and the electrons. Since the ionosphere is neutral we can set,

$$N_i = N_e = N \tag{2.3-1}$$

Let E_x be the electric field of the incoming wave, and let us consider its effects inside the pathway of the beam from the radio transmitter. Let dx be a small segment of the plasma along the x-axis inside the restricted region. Under the $-eE$ force of the electric field the electrons will move a distance ξ in the opposite direction of the field and their volume will change by $d\xi$ in the x-direction (Figure 2.3-I). The new electron density N_e' will be,

$$N_e' = N_e\left(\frac{dx}{dx - d\xi}\right) = N\left(\frac{1}{1 - \dfrac{d\xi}{dx}}\right) = N\left(1 + \frac{d\xi}{dx}\right) = N + N\left(\frac{d\xi}{dx}\right) \tag{2.3-2}$$

The initial charge density ϱ_C was equal to zero,

$$\varrho_C = eN_i - eN_e = 0 \tag{2.3-3}$$

but after the electron density has changed, while the ion density has remained the same ($N'_i = N_i$), the new charge density ϱ'_c will be,

$$\varrho'_c = eN'_i - eN'_e = eN - eN - eN\left(\frac{d\xi}{dx}\right) = -eN\left(\frac{d\xi}{dx}\right) \qquad (2.3\text{-}4)$$

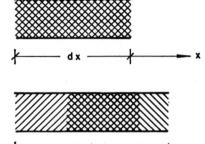

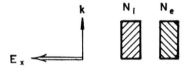

FIGURE 2.3-I   The motion of the electrons past the quasistationary ions due to a weak external electric field E_x

From Maxwell's equations we have $\nabla \cdot E = 4\pi\varrho_c$. In the present case, where the electric field is in the $-x$ direction, we have $\nabla \cdot E = -(dE_x)/dx$ and using the charge density we obtained in (2.3-4) we get,

$$\nabla \cdot E = \frac{-dE_x}{dx} = 4\pi\left(-eN\frac{d\xi}{dx}\right) = 4\pi\varrho_c \qquad (2.3\text{-}5)$$

and therefore,

$$dE_x = 4\pi eN\, d\xi \qquad (2.3\text{-}6)$$

Integrating (2.3-6) and setting the integration constant equal to zero ($\xi = 0$ when $E_x = 0$) we obtain,

$$E_x = 4\pi eN\xi \qquad (2.3\text{-}7)$$

The separation of a positive and a negative charge produces a restoring Coulomb force F_x given by the expression,

$$F_x = -eE_x = -4\pi e^2 N\xi \qquad (2.3\text{-}8)$$

which, according to Newton's law, must be equal to the mass times the acceleration of the electrons since the electrons will essentially rush back toward the much heavier ions. Hence we have,

$$-4\pi e^2 N\xi = m\left(\frac{d^2\xi}{dt^2}\right) \tag{2.3-9}$$

This, however, is the equation of the harmonic oscillator which physically means that the electron, having picked up enough kinetic energy due to the force F_x, will overshoot the position of the ion. Then the force F_x will bring it back again, but again it will go beyond the position of equilibrium and thus it will remain in a harmonic motion about the position of the positive ion. The solution of (2.3-9) is,

$$\xi = \xi_0 e^{-i\omega_N t} \tag{2.3-10}$$

where ω_N is the *angular plasma frequency* of the medium given by the expression,

$$\omega_N = \left(\frac{4\pi e^2 N}{m}\right)^{1/2} \tag{2.3-11}$$

Finally, $f_N = \omega_N/2\pi$ is the *plasma frequency* of the medium and is given by the simple formula,

$$f_N = \left(\frac{e^2 N}{\pi m}\right)^{1/2} = 9 \times 10^3 \, N^{1/2} \tag{2.3-12}$$

In the numerical form of (2.3-12), f_N is measured in Hz and N in electrons per cm³. Thus when $N = 10^6$ el/cm³, $f_N = 9$ MHz.

In conclusion, the plasma frequency of an ionized region is the natural frequency at which the electrons of the region would oscillate about their position of equilibrium if their original condition was disturbed. The disturbance in this case is caused by the electric field of the wave which also varies in a harmonic fashion with the frequency f of the radio wave, i.e.,

$$E_x = E_0 \cos \omega t = E_0 \cos 2\pi f t \tag{2.3-13}$$

As a result, the electrons become forced harmonic oscillators because they are forced to oscillate in the frequency of the radio wave rather than in their own natural plasma frequency. The equation of the forced harmonic oscillator with an external force $F_0 \cos \omega t$ is;

$$m\left(\frac{d^2\xi}{dt^2}\right) = -m\omega_N^2 \, \xi + F_0 \cos \omega t \tag{2.3-14}$$

and its solution is,

$$\xi = \xi_0 \cos \omega t \tag{2.3-15}$$

where the amplitude of the oscillations ξ_0 is given by the expression,

$$\xi_0 = \frac{F_0}{m(\omega^2 - \omega_N^2)} \qquad (2.3\text{-}16)$$

When the two frequencies are far apart, the amplitude ξ_0 is small and tends to zero for very large values of ω. When on the other hand ω approaches ω_N, the amplitude of the oscillations becomes very large. It is very much like pushing a child on a swing. One gets the best results when the periodic pushes are coordinated with the natural period of the swing. At $\omega = \omega_N$ the amplitude appears to become infinite, but this does not actually happen because frictional and other forces that are normally negligible become important near the resonance frequency.

As the radio waves pass through a plasma, the electrons absorb energy from the radio waves for their oscillations and, as all harmonic oscillators do, they reradiate it in the frequency of their forced oscillations, i.e., at the frequency of the radio waves. The reradiated waves either reinforce or cancel the passing radio waves, depending on their relative phase. At very large frequencies the amplitude of the reradiated waves becomes vanishingly small and the radio waves pass through the plasma practically undisturbed. As ω approaches ω_N, the reradiated waves become stronger and tend to cancel the passing radio waves in the forward direction. At the same time a substantial amount of energy is reradiated in the backward direction. At frequencies below the plasma frequency of the medium the transmitted (forward) wave tends to zero, while the reflected (backward) wave tends to reach the full intensity of the incoming wave.

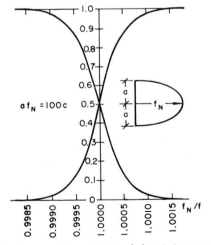

FIGURE 2.3-II   The change of the reflection $|R|$ and the transmission $|T|$ coefficients with frequency, for a parabolic electron density profile

In the full wave solution of the problem, like in quantum mechanics, the transmission and reflection coefficients vary smoothly with frequency from 0 to 1. As seen from Figure 2.1-II, which describes the passage of radio waves through a parabolic layer of maximum plasma frequency f_N, most of this variation takes place very near f_N and for this reason we can adopt the classical step function formulation and simply state that radio waves with $f > f_N$ will be able to pass through this medium whereas radio waves with $f < f_N$ will be reflected.

The group velocity of radio waves, i.e., the velocity with which a group of radio waves (a radio signal) propagates through a plasma, is given by the relation,

$$V_{gr} = \frac{c}{\mu_{gr}}$$ (2.3-17)

where μ_{gr} is the *group index of refraction*, which is related to the *index of refraction* μ of the plasma through the expression,

$$\frac{1}{\mu_{gr}} = \mu = \left(1 - \frac{f_N}{f^2}\right)^{1/2}$$ (2.3-18)

From (2.3-17) and (2.3-18) it follows that $V_{gr} = \mu c$, which says that the group velocity becomes zero when $\mu = 0$, i.e., when,

$$f = f_N = \left(\frac{e^2 N}{\pi m}\right)^{1/2}$$ (2.3-19)

which occurs, as we have seen, when the waves are about to be reflected. It should be made clear that this is the case only for normal incidence. When the radio waves approach a plasma layer at an angle θ to the normal, then the critical frequency (the highest frequency reflected by the layer) f_c is,

$$f_c = f_N \sec \theta$$ (2.3-20)

In accordance with what we have discussed up to now, radio waves transmitted vertically from the ground will be reflected in the ionosphere at a height where the plasma frequency of the ionosphere becomes equal to the frequency of the wave. As seen from (2.3-20), for oblique transmission the same layer will be able to reflect considerably higher frequencies. Waves of higher frequencies will be reflected at higher layers, i.e., closer to the maximum of the ionosphere. Multiplying the time of travel of the reflected radio pulses with the speed of light, we can find the height of the layer at which the pulse was reflected. This, however, is not exactly correct because the pulses as we have seen in (2.3-17) travel with a velocity $V_{gr} < c$ and therefore the height obtained in the above manner is an overestimate. For this reason, it is called the *equivalent height* and is usually denoted by h'.

The presence of the earth's magnetic field makes the ionosphere a *magnetoactive plasma*, i.e., a plasma with an embedded magnetic field. Radio waves in magnetoactive plasmas split into two modes of propagation called the *ordinary* and the *extraordinary*. Each mode has its own index of refraction which is much more complicated than (2.3-18). As a result the two modes propagate with different group velocities and are reflected at different heights in the ionosphere. Thus for each transmitted radio pulse we receive back two separate echoes. This is clearly seen in the *ionogram* of Figure 2.3-III. The horizontal axis of the ionogram gives the transmission frequency of the ionospheric sounder, and the vertical axis is the equivalent height. To a first approximation, the ordinary f_0 and the extraordinary f_x frequencies reflected from the same ionospheric layer are related through the expression,

$$f_x - f_0 = \frac{1}{2} f_H \qquad (2.3\text{-}21)$$

where f_H is the *cyclotron frequency* of the earth's magnetic field H.

$$f_H = \frac{1}{2\pi} \frac{eH}{mc} \qquad (2.3\text{-}22)$$

If H is expressed in Gauss and f_H in MHz, then $f_H = 2.8H$. In the terrestrial ionosphere $f_H \simeq 1.0 - 1.5\,\text{MHz}$. As seen from (2.3-21) the highest frequency that will be reflected by the ionosphere will be the frequency of the extraordinary mode reflected at N_{max}. This is called the *maximum usable frequency* (MUF) and its values and variations around the globe are of great importance to all radio telecommunications.

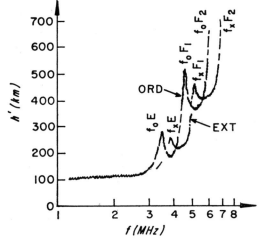

FIGURE 2.3-III A typical ionogram showing the ordinary and the extraordinary traces from the different ionospheric layers

2.4 Collision Frequency and Absorption

The rate at which electrons collide with neutral particles and ions is called *collision frequency*. If a passing radio wave had set the electrons in harmonic motion, these collisions would disrupt it and the ordered (harmonic) energy of the electrons will be converted into random (thermal) kinetic energy. As a result, the radio wave will have to spend some more of its energy to start again the harmonic motion of the electrons and in this manner collisions cause the attenuation of radio waves. Collisions of electrons with electrons, because both particles have the same mass, contribute much less to the thermalization of their energy than do collisions with the much heavier ions and neutral particles. For this reason, electron-electron collisions can be neglected in most cases in computing the collision frequency which causes the attenuation of the passing radio waves.

The collision frequency of electrons with the neutral particles v_n is proportional to the physical cross-section of the neutral particles $\sigma_n \simeq 10^{-15}\,\text{cm}^2$, their concentration N_n, and to the thermal velocity of the electrons V_e. Thus we have,

$$v_n = \sigma_n N_n V_e = (\pi r_n^2)\, N_n \left(\frac{3kT}{m}\right)^{1/2} \simeq C_n N_n T^{1/2} \qquad (2.4\text{-}1)$$

where C_n is a numerical constant of the order of 10^{-10} in the CGS system. The collisions of electrons with ions are actually Coulomb collisions in which the electrons are scattered by the ions through the interaction of their electric fields rather than through physical contact. The collision cross-section, therefore, is much larger than the physical cross-section of the ions. As a first approximation one can say that the maximum distance for an effective interaction is the distance r at which the kinetic energy of the electrons is equal to the Coulomb potential of the two particles, i.e.,

$$\frac{e^2}{r} = \frac{1}{2} m V_e^2 = \frac{3}{2} kT \qquad (2.4\text{-}2)$$

Using (2.4-2) we find that the collision frequency of the electrons with ions v_i is,

$$v_i = \sigma_i N_i V_e = \pi \left(\frac{2e^2}{3kT}\right)^2 N_i \left(\frac{3kT}{m}\right)^{1/2} = C_i N_i T^{-3/2} \qquad (2.4\text{-}3)$$

where C_i is a numerical constant of the order of 10 in the CGS system. Finally, using the fact that $N_i = N_e = N$, we can write the total collision frequency in the form,

$$v = v_n + v_i = C_n N_n T^{1/2} + C_i N T^{-3/2} \qquad (2.4\text{-}4)$$

When the Debye shielding of the ions is taken into consideration, C_i is not merely a constant but it includes also a logarithmic term of the temperature and the electron density. Several authors have presented values of these constants in the literature. The following expression is the one obtained by Nicolet (1953) for the total collision frequency,

$$\nu = 5.4 \times 10^{-10} N_n T^{1/2} + \left\{ 34 + 4.18 \log_{10}\left(\frac{T^3}{N}\right) \right\} NT^{-3/2} \qquad (2.4\text{-}5)$$

From the collision frequency we can now compute the absorption coefficient. It can be shown that the damping force due to collisions is to a first approximation proportional to the velocity $V = \dot{r}$ of the electrons. Also that the constant of proportionality is equal to $m\nu$, where m is the mass of the electron and ν the collision frequency. Hence the equation of motion of an electron under the oscillating force of a field E and in the presence of collisions is,

$$m\ddot{r} + m\nu\dot{r} = -eE \qquad (2.4\text{-}6)$$

Since $E = E_0 e^{i\omega t}$ and $r = r_0 e^{i\omega t}$, we can write (2.4-6) in the form,

$$-m\omega^2 r + im\nu\omega r = -eE \qquad (2.4\text{-}7)$$

Now, since the dipole moment between an electron and an ion separated by a distance r is equal to $-er$, and since we have N electron-ion pairs per unit volume, the *polarizibility* P, i.e., the induced dipole moment per unit volume is $P = -Ner$. By introducing P in (2.4-7) we obtain,

$$\frac{m\omega^2}{eN} P - \frac{im\nu\omega}{eN} P = -eE \qquad (2.4\text{-}8)$$

which then gives,

$$P = -\left[\frac{e^2 N}{m\omega^2 \left(1 - i\dfrac{\nu}{\omega}\right)} \right] E = -\left[\frac{\omega_N^2}{4\pi\omega^2 \left(1 - i\dfrac{\nu}{\omega}\right)} \right] E \qquad (2.4\text{-}9)$$

where ω_N is the angular plasma frequency given by (2.3-11). From the electromagnetic theory on the other hand we have,

$$D = \varepsilon E = E + 4\pi P = E - \frac{\omega_N^2/\omega^2}{\left(1 - i\dfrac{\nu}{\omega}\right)} E = \left[1 - \frac{\omega_N^2/\omega^2}{\left(1 - i\dfrac{\nu}{\omega}\right)} \right] E$$

$$(2.4\text{-}10)$$

from which we conclude that the square of the *complex index of refraction* n^2, which is equal to the *complex dielectric constant* ε, is given by the expression,

$$n^2 = \varepsilon = 1 - \frac{\omega_N^2/\omega^2}{\left(1 - i\dfrac{v}{\omega}\right)} = \left[1 - \frac{\omega_N^2/\omega^2}{1 + \left(\dfrac{v}{\omega}\right)^2}\right] - i\left[\left(\frac{v}{\omega}\right)\frac{\omega_N^2/\omega^2}{1 + \left(\dfrac{v}{\omega}\right)^2}\right]$$

(2.4-11)

Let μ then be the real part and χ the imaginary part of the complex index of refraction so that,

$$n = \mu - i\chi \tag{2.4-12}$$

From (2.4-11) and (2.4-12) it follows that,

$$\mu^2 - \chi^2 = 1 - \frac{\omega_N^2/\omega^2}{1 + \left(\dfrac{v}{\omega}\right)^2} \tag{2.4-13}$$

$$2\mu\chi = \left(\frac{v}{\omega}\right)\frac{\omega_N^2/\omega^2}{1 + \left(\dfrac{v}{\omega}\right)^2} \tag{2.4-14}$$

When the collision frequency is much smaller than the operating frequency, i.e. when $v \ll \omega$, then $\chi \ll 1$ and the above equations become,

$$\mu^2 \simeq 1 - \frac{\omega_N^2}{\omega^2} \tag{2.4-15}$$

and,

$$\chi \simeq \frac{v\omega_N^2}{2\mu\omega^3} \tag{2.4-16}$$

The electric field of a wave in a medium with a complex index of refraction n is given by the expression,

$$E = Ae^{-i(kr-\omega t)} = Ae^{-i(nk_0 r - \omega t)}$$

$$= Ae^{-i(\mu k_0 r - \omega t)}e^{-k_0 \chi r} = E_0 e^{-k_0 \chi r} \tag{2.4-17}$$

and since the intensity of the radiation I is proportional to the square of the electric field E we have,

$$I = I_0 e^{-2k_0 \chi r} = I_0 e^{-\varkappa r} = I_0 e^{-\tau} \tag{2.4-18}$$

where $\varkappa$ is the *absorption coefficient* of the medium,

$$\varkappa = 2k_0\chi = 2\frac{\omega}{c}\chi = 2\frac{\omega}{c}\frac{v\omega_N^2}{2\mu\omega^3} = \frac{v}{\mu c}\frac{f_N^2}{f^2} \tag{2.4-19}$$

and τ the *optical thickness* or *opacity* of the medium,

$$\tau = \int \varkappa \, dr \simeq \varkappa r \qquad (2.4\text{-}20)$$

From (2.4-19) and (2.4-20) we see that the absorption coefficient $\varkappa$ and the opacity τ of the medium are directly proportional to the collision frequency ν which is given by (2.4-5). For a very weakly ionized plasma, like the lower ionosphere (*D*-region) where $N_n \gg N_i$, we can set $\nu = \nu_n$ and by introducing (2.4-1) and (2.3-12) in (2.4-19) we get,

$$\varkappa_n \propto \frac{N N_n T^{1/2}}{f^2} \qquad (2.4\text{-}21)$$

The electron density N in the *D*-region is usually low and therefore the $\varkappa_n$ of the *D* region is usually low. Occasionally, however, N increases by a factor of 10 or 100 and with the very large number of neutral particles N_n available makes the $\varkappa_n$ of the *D*-region temporarily very large. The absorption of the *D*-layer is usually measured with radio receivers which monitor continuously the radio noise from our galaxy. These receivers are called *riometers,* where the prefix *rio* stands for the initials of the words *relative ionospheric opacity*. During the span of a day both the ionospheric absorption and the galactic radio background change. The first because the electron density of the *D*-layer varies with the zenith angle of the sun, and second because the galactic radio noise is concentrated in the plane of the galaxy and especially toward the galactic center and as the earth rotates our antenna focuses on different regions of the galaxy. Through long observations, however, we can take into account these variations, which have also a seasonal component, and we can establish the normal levels of ionospheric absorption and the expected, under normal conditions, intensity I of the galactic radio emission. Any decrease of the signal strength to a new level I' represents an increase in the *D*-region absorption and is usually expressed in *db* units,

$$db = 10 \log \left(\frac{I'}{I}\right) = 23 \ln \left(\frac{I_0 e^{-\tau'}}{I_0 e^{-\tau}}\right) = 23 \ln e^{(\tau - \tau')}$$

$$= -23 \, \tau' \left(1 - \frac{\tau}{\tau'}\right) \simeq -23 \, \tau' \qquad (2.4\text{-}22)$$

where τ and τ' are the normal and the enhanced opacity of the ionosphere, which are essentially proportional to the normal and enhanced electron density of the *D*-region. Riometer observations are usually conducted in the frequency range between 15 and 60 MHz. The reason is that lower frequencies can hardly penetrate through the ionosphere while higher frequencies, as seen from (2.4-19), suffer very little attenuation which is difficult to measure.

For a fully ionized plasma, like the solar corona, we can set $\nu = \nu_i$ and by introducing (2.4-3) in (2.4-19) and (2.4-20) we get,

$$\tau_i = \varkappa_i r \propto \left(\frac{N^2}{f^2 T^{3/2}} \right) r \qquad (2.4\text{-}23)$$

Here it is important to note that in certain cases though $\varkappa_i$ might be very small, the total attenuation represented by the opacity τ_i might be quite large simply because the radio waves have travelled a very long distance r in the medium. This is quite often the case in astronomical observations.

2.5   The Structure of the Ionosphere and the Plasmasphere

The Chapman electron density profile is only a first approximation to the actual structure of the ionosphere. A typical profile of the ionosphere is shown in Figure 2.5-I. The most obvious difference from the simple Chapman layer theory is that the ionosphere has several peaks of the electron density. These peaks are called *layers* and have the names *D-layer, E-layer, F1-layer,* and *F2-layer*. We also have the *plasmapause*, a rather sharp change in electron density, at a distance of 4 to 5 earth radii. The D, E, $F1$ and $F2$ layers are formed because the ionizing radiation from the sun is not mono-chromatic and because the atmosphere consists of several different consti-

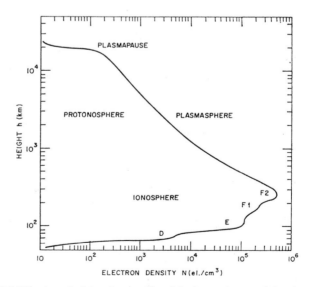

FIGURE 2.5-I   A typical daytime profile of the ionosphere and the plasmasphere

tuents which are ionized at different wavelengths of the solar spectrum. Each one of these ionizing processes reaches its peak at a different altitude which becomes an individual peak of the ionospheric electron density profile. The height dependence of the different loss-mechanisms is also a factor in the formation of the ionospheric layers. It should be mentioned that the valleys between the maxima of the electron density are in general quite shallow, and therefore it is only for historical reasons that we still occasionally call "layer" the region surrounding a given peak. In general we recognize the following layers, or better, regions of the ionosphere.

The D-region It is present only during the day and covers the range between 60 and 85 km. Quite often the valley between the *D*-layer and the *E*-layer is not very obvious, and as a result the *D*-layer is not always well defined. Below 70 km cosmic rays are the principal source of ionization. These highly energetic particles penetrate deep into the atmosphere and the electrons they produce at these low altitudes are quickly lost by attaching themselves to neutral particles to form *negative ions*. These electrons are liberated during the day through *photodetachment* by the solar radiation while during the night they remain bound to the neutral atoms. Thus, though the cosmic ray flux remains essentially unchanged around the clock, the above mentioned mechanism explains the night to day differences observed in the electron density of the lower *D*-region. In the 70 to 85 km range electrons are produced mainly from the ionization of the traces of NO that exist in the upper atmosphere by the Lyman-α radiation (1216 Å) of the sun. The peak electron density of the *D*-layer occurs near 80 km and is of the order of 3×10^3 el/cm^3. Solar x-rays acting on molecular oxygen and molecular nitrogen contribute also to the ionization of the *D*-layer. This becomes especially apparent during periods of intense solar activity (solar flares, etc.) when the electron content of the *D*-region can increase by more than an order of magnitude. During such periods, as we have seen at the end of Section 2.4, the absorption of the *D*-layer shows a marked increase with serious repercussions for radio telecommunications.

The E-region It extends from 85 km to about 150 km and has a daytime maximum of $\sim 10^5$ el/cm^3 around 115 km. During the night the electron density decreases by at least two orders of magnitude and the *E*-layer disappears. The lower part of the E-region (85–100 km) is ionized mainly by solar x-rays in the 30–100 Å range. Above 100 km the ionization is produced mainly by soft x-rays and by ultraviolet radiation in the range between 800 Å and the Lyman-β at 1026 Å. It should be noted, however, that there ave still a lot of uncertainties in the photoionization of the ionosphere (Allen, 1965). The main ions in the *E*-region are NO$^+$ and O$_2^+$.

Though N_2^+ is produced in large numbers it is virtually absent in the ionosphere because of its extremely high recombination rate. The recombination coefficient (2.2-10) of the E-layer is,

$$\alpha \simeq 2 \times 10^{-8} \text{ cm}^3 \text{ sec}^{-1}$$

and the relaxation time (2.5-2) is of the order of 10 minutes, which explains the rapid disappearance of the E-layer after sunset. The small amount of ionization which persists in the E-region during the night could be due to micrometeorite bombardment. The *relaxation time* t_r is the time in which the electron density would reduce to one half if there was no more production of electrons. Setting $q = 0$ in (2.2-13) and neglecting attachment and diffusion we get,

$$\frac{dN}{dt} = \alpha N^2 \tag{2.5-1}$$

Solving this simple differential equation we find that the time t_r in which $N(t_r)$ becomes equal to $\frac{1}{2} N(0)$ is given by the expression,

$$t_r = \frac{1}{\alpha N(0)} \tag{2.5-2}$$

Studies for the experimental determination of t_r can best be made during solar eclipses. In general the E-layer behaves very much like a type-α Chapman layer to which one can assign a scale height $H \simeq 10$ km.

The F1-region It is present like the D- and E-layers only during the day. It extends from 150 to 200 km with a typical maximum of 2×10^5 el/cm^3 around 180 km. The principal ionizing agent is the sun's ultraviolet radiation in the 200 Å to 900 Å range. The main atmospheric constituent which is ionized in the $F1$-region is atomic oxygen which diffuses from the lower layers where it is produced from the dissociation of O_2. The photoionization of the atmosphere by the ionizing solar radiation produces electrons with relatively high kinetic energies. These electrons, which are called *photoelectrons* because of the mode of their production, share later their energy with the other ambient electrons and thus raise the kinetic temperature of the electron gas several hundred degrees above the kinetic temperature of the neutral gas. A few of the photoelectrons that are produced at higher altitudes, where collisions are relatively rare, can travel along a magnetic field line to the opposite hemisphere (conjugate point) where they also heat up the electron gas. This effect has been observed very clearly when the time of sunrise differs by one or two hours at the two conjugate points (Carlson, 1966). The recombination coefficient of the $F1$-layer is $\alpha \simeq 5 \times 10^{-9}$ cm^3 sec^{-1} and the corresponding relaxation time is similar to the relaxation time of the E-layer.

The F2-region It extends from 200 km to roughly 1000 km and has a daytime maximum near 250 km of about 5×10^5 el/cm^3. During the night the D, E, and F1 peaks disappear and the ionosphere takes the form of a single layer, called the F-layer, with a maximum of about 10^5 el/cm^3 in the vicinity of 350 km. As we move to higher altitudes the rate at which electrons and ions recombine decreases rapidly with height and the relaxation time at higher altitudes is much longer. In the upper F2-region, for example the relaxation time is of the order of several days, which makes it possible for the earth to retain a substantial ionosphere during the night.

The variation of the loss-processes with height is the cause for the appearance of the F2 peak. The F1- and the F2-regions have the same electron production profile $q(h)$ and the F1 peak corresponds to the maximum q_m of the electron production profile. The loss mechanisms, however, are different in the two regions and the F2-layer is formed simply because the loss rate at this altitude is much smaller. Under quasi steady-state conditions, i.e. when $dN/dt \simeq 0$, the continuity equation (2.2-13) becomes,

$$q = L \qquad (2.5\text{-}3)$$

Figure 2.5-II shows in a schematic way the change of q, L, and N with height. It is easily seen from Figure 2.5-II that the F2 peak is formed because the loss rate decreases rapidly as the dominant term in L changes from αN^2 to βN in the F2-region.

Electrons can be lost by recombining directly with the positive ions, which are present in approximately equal numbers. When two particles, however, combine to form a single particle it is impossible to concerve both energy and momentum. For this reason this reaction can take place only in the presence of a neutral particle in a three-body collision. In the D-region, where neutral particles are extremely plentiful, this process works well, but in the E and especially in the F-region three-body collisions become rare and this mechanism is replaced by a loss process which requires only two-body collisions. In this process, the ions of atomic oxygen, which is the principal ionizable gas of these heights, pass their charge to the neutral molecules of the atmosphere,

$$O^+ + O_2 = O + O_2^+ \qquad \text{(charge transfer)}$$

or

$$O^+ + N_2 = N + NO^+ \qquad \text{(ion-atom interchange)} \qquad (2.5\text{-}4)$$

and then the new ions are dissociated in neutralizing collisions with the free electrons,

$$e^- + O_2^+ = O + O$$

or

$$e^- + NO^+ = N + O \qquad \text{(dissociative recombination)} \qquad (2.5\text{-}5)$$

Note that both (2.5-4) and (2.5-5) are two-body collisions which produce also two particles, thus satisfying the laws of conservation of energy and momentum. It should be mentioned that the neutral atoms of (2.5-5) are occasionally produced in excited states and as they return to their ground state they emit photons which are responsible for a faint glowing of the sky, a phenomenon which is called *airglow*.

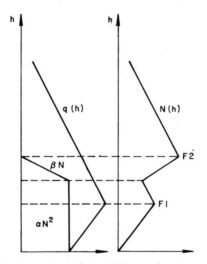

FIGURE 2.5-II   The formation of the $F2$-layer due to a change in the loss mechanism with height. The production of ionization q peaks at the $F1$-layer

In the E and $F1$ regions, neutral particles are still plentiful enough and (2.5-4) proceeds much faster than (2.5-5), and therefore (2.5-5) becomes the loss rate determining step. From (2.5-5) it follows that the loss rate is proportional to the number density of the electrons N and the number density of the molecular ions. But since (2.5-4) is in equilibrium, the latter is proportional to the number density of the atomic ions, which is essentially equal to the electron density N. Hence we have $L_R = \alpha N^2$. In the $F2$-region, however, where the number density of neutral molecules is considerably smaller, (2.5-4) becomes the bottleneck reaction. Thus the loss rate in the $F2$-region is proportional only to the ion density, i.e., to the electron density $N(L_A = \beta N)$, and has the form of an attachment-like loss process (as if the electrons were lost by becoming attached to the more plentiful neutral molecules). For this reason the constant of proportionality β, is called the *attachment coefficient* and is a function of the concentration of neutral molecules. Since the density of neutral particles decreases with height, β decreases also with height. Ratcliffe et al. (1956) have derived the following semi-empirical relation for the change of the attachment coefficient β with altitude,

$$\beta = 10^{-4} \exp\left(\frac{300 - h}{50}\right) \sec^{-1} \tag{2.5-6}$$

From this relation it follows that the relaxation time ($t_r = \ln 2/\beta$) is of the order of 5 hours at 350 km and of the order of 40 hours at 500 km. This explains how the earth manages to retain a substantial ionosphere during the night and why the nighttime peak of the F-region occurs at higher altitudes.

The Upper Ionosphere At altitudes above the $F2$ peak both the production and the loss of electrons tend to zero, and the only term left in the right hand side of (2.5-13) is the term $\nabla \cdot (NV)$, which means that the upper ionosphere is maintained through the upward diffusion of ionization. In the presence of the earth's magnetic field, which tends to guide the diffusion of the charged particles along the field lines, this becomes a very complicated phenomenon to study. Contrary to the diffusion of neutral particles, where different species diffuse independently, electrons and ions tend to diffuse as a body because their separation builds up an electric field which brings them back together. This phenomenon is called *ambipolar diffusion*.

Neglecting drag forces and the effects of the earth's magnetic field, we can obtain two simple equations for the hydrostatic equilibrium of the electrons and the ions under the force of gravity and the electric field between the two species. Assuming an isothermal upper ionosphere in which, however, the temperatures of the ions and electrons are not necessarily equal we have,

$$\frac{dP_e}{dh} = kT_e\left(\frac{dN_e}{dh}\right) = m_e N_e g - eN_e E \tag{2.5-7}$$

$$\frac{dP_i}{dh} = kT_i\left(\frac{dN_i}{dh}\right) = m_i N_i g + eN_i E \tag{2.5-8}$$

Since the ionosphere remains neutral, we have that $N_i = N_e = N$. Adding (2.5-7) and (2.5-8) we obtain,

$$k(T_e + T_i)\frac{dN}{dh} = -(m_e + m_i)gN \tag{2.5-9}$$

Finally by neglecting m_e with respect to m_i we can write (2.5-9) in the form,

$$\frac{d}{dh}\ln N = -\frac{m_i g}{k(T_e + T_i)} \tag{2.5-10}$$

4*

Equation (2.5-10) shows that if $T_e = T_i$, the upper ionosphere will be in hydrostatic-like equilibrium with a scale height approximately twice the scale height of the neutral atmosphere. Of course, there are several different ions which tend also to be differentiated by height. Thus around 1000 km O^+ is replaced by He^+ as the predominant ion, and at even higher altitudes ($\sim 2,500$ km) He^+ is replaced by H^+, i.e., by free protons. The layer where helium ions dominate is often called the *heliosphere* and the region above it is called the *protonosphere*.

The Plasmasphere   This is the region of the earth's ionized atmosphere which basically follows the rotation of the earth. The plasmasphere has the shape of a doughnut, very much like the volume formed by the lines of the earth's dipole magnetic field which provides the link that keeps the plasmasphere rotating with the earth. The boundary of the plasmasphere, which at the equatorial plane occurs at a geocentric distance of 4 to 5 earth radii, is called the *plasmapause*. At the plasmapause the electron density drops sharply from a few hundred el/cm³ to only a few el/cm³. The plasmasphere is filled with thermal plasma (a plasma with a Maxwellian distribution of velocities and a temperature of a few thousand degrees Kelvin) which diffuses upwards from the upper ionosphere.

The reverse process, i.e. a downward flux of ionization, seems to occur during the night and in general there is a strong coupling between the upper ionosphere and the plasmasphere. This can become very important under special conditions, such as during geomagnetic storms, when we might observe a very pronounced downward flux of ionization from the plasma-

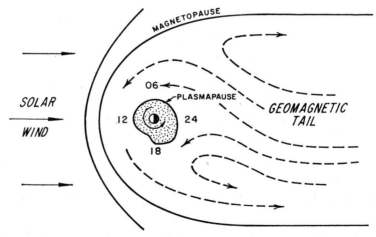

FIGURE 2.5-III   The formation of the plasmapause due to the interaction of the convective stream from the geomagnetic tail with the co-rotating plasmasphere

sphere to the *F*-region. Beyond the plasmapause the density of the thermal plasma drops to zero but this is not an empty space because the region between the plasmapause and the magnetopause is filled with energetic electrons and protons. Streams of such energetic particles, most likely of solar wind origin, flow continuously in and out of this region from the geomagnetic tail. A certain asymmetry in the flow pattern (Figure 2.5-III) is due to the rotation of the earth's magnetic field with the earth. The driving force for this convection is an equatorial electric field along the sunrise to sunset line which is produced by the interaction of the solar wind with the magnetosphere of the earth.

Satellite observations and especially whistler measurements by Carpenter (1966) (whistlers are radio noise signals in the kHz range that travel along the field lines of the earth's magnetic field and yield information on the electron densities at the equatorial crossing of the lines) have shown that the equatorial cross-section of the plasmapause differs considerably from a perfect circle. As seen from Figure 2.5-III, the plasmapause develops a sudden bulge near the sunset meridian where its radial distance increases from about 3.5 to nearly 5 earth radii. The bulge persists through the 18:00 to 21:00 hour sector and finally subsides near the mignight meridian. Brice (1967) and other authors who have studied this phenomenon suggest that there is a continuous convection of plasma from the geomagnetic tail which interacts with the plasmasphere that co-rotates with the earth. In the sun-rise sector (06:00 hr) the directions of the two motions coincide. But in the sun-set sector (18:00 hr) the convective plasma opposes the motion of the co-rotating plasma and produces a pile-up in the plasmasphere which forms the bulge. Figure 2.5-III is a diagrammatic representation of the convective streams and the plasmaspheric rotation in the equatorial plane.

The convective flux from the tail increases considerably with geomagnetic activity and the earth's magnetic field can carry along only a smaller volume of plasma against the braking action of the enhanced convection. As a result, during large magnetic storms the radial distance of the plasmapause decreases by about one earth radius and in general there is a very close correlation between the radial distance of the plasmapause, especially at the sensitive dusk sector, and geomagnetic activity (Mendillo and Papagiannis, 1971).

2.6 Regular and Irregular Variations of the Ionosphere

The ionosphere we have described up to now and the numerical values we have given refer to an average, or typical as some people prefer to call it, ionosphere. In practice these values vary by more than an order of magnitude with time and location. Some of these changes follow a known pattern, whereas others come and go on an irregular basis. The regular variations include:

The Latitudinal Dependence of the ionospheric parameters, mainly due to the change of the solar zenith angle with latitude, but also due to the change in the dip angle of the earth's magnetic field. There is also a small longitudinal variation because the earth's magnetic field varies with longitude along any given geographic latitude. The N_m can easily vary by an order of magnitude from the polar to the equatorial regions.

The Diurnal Variation of the ionosphere which includes the peaking of the electron density usually in the early afternoon, the sharp changes near sunrise and sunset, and the disappearance of the lower layers during the night. The N_m can again vary by an order of magnitude between night and day.

The Seasonal Variation, which is also due to the change in the average zenith angle of the sun as we move between the summer and winter solstices.

The 27 Day Cycle due to the intrinsic rotation of the sun. This cycle is especially noticeable during periods of high solar activity when a very active region might last for more than one rotation of the sun. Active regions also have a tendency to form in the same general area of other past active regions so that there is often a long lasting longitudinal asymmetry of activity on the sun (Sawyer, 1968).

The 11 Year Solar Cycle, which represents the fairly regular increase and decrease of the solar activity and therefore of the ionizing radiation from the sun with a period of approximately 11.2 years. The last solar maximum occurred in 1969.

The fact that all these variations follow a rather well-prescribed pattern does not necessarily mean that these patterns follow the predictions of the simple Chapman layer theory. According to the Chapman theory, for example, the highest $f_0 F2$ and the lowest h_m must occur when the sun reaches the smallest zenith angle, which naturally occurs at noon. The Chapman theory also predicts lower critical frequencies at higher latitudes and for the same latitude lower critical frequencies in the winter hemispere. All the variations of the ionosphere that do not follow the predictions of the Chapman theory came to be known as *anomalies* and over the years many anomalies of this kind have been reported and discussed in the literature. Thus we have:

The Equatorial or Geomagnetic Anomaly, because the f_0F2 varies with the geomagnetic rather than with the geographic latitude plus the fact that the noon values of the f_0F2 show a decrease (*trough*) along the geomagnetic equator at the equinoxes.

The Seasonal Anomaly, because at mid-latitudes the noon values of the f_0F2 are higher in the winter than in the summer hemisphere.

The December Anomaly, because on a world-wide basis the f_0F2 values of the ionosphere are in general higher around December.

The Diurnal Anomaly, because the diurnal variation of the f_0F2 is not always symmetric around the local noon. This phenomenon is especially pronounced at mid-latitudes during the summer months when the evening and early night values of the f_0F2 approach and often exceed the corresponding noon value. The segment of high critical frequencies around noontime which is missing from the diurnal plot of the f_0F2 has been given the descriptive name *midday bite-out*. A worldwide study of the high post-sunset values of the f_0F2 by Papagiannis and Mullaney (1971) has revealed that the geographical distribution of the magnitude of this anomaly is closely coupled to the configuration of the earth's magnetic field and the global pattern of the neutral winds that blow at ionospheric heights.

People have tried to account for these so-called anomalies by including effects that were neglected by the simple Chapman theory. Some of the most important ones are:

1   Ambipolar diffusion in the presence of the earth's magnetic field.

2   The coupling between the ionosphere and the plasmasphere.

3   The dragging of ionospheric plasma by neutral winds in the upper atmosphere.

4   The change with temperature of the production rate, the loss rate, and the scale height at any given altitude.

People have also investigated different conjugate point effects such as the transport of ionization and the transport of energy by photoelectrons between the two conjugate hemispheres along the lines of the earth's magnetic field. A complete theory for the $F2$-region, which essentially has no upper boundary, is very difficult because all the effects mentioned above should be taken into account. It is again the ingenuity of the investigator to recognize the most important of these effects in order to build a manageable model which can overcome the deficiencies of the simple Chapman theory.

Besides the different anomalies which we have discussed above, the ionosphere shows also the following structural irregularities:

The Sporadic-E, which is the frequent formation of a thin layer (1–5 km) of
excess ionization at an altitude of about 110 km. The electron density of this
layer can exceed by more than a factor of two the ambient electron density
of the *E*-region. The sporadic-*E* has been studied extensively both from
the theoretical and the experimental point of view, but still there is no general
agreement on the cause of this phenomenon. According to one of the more
widely discussed theories, the appearance of the sporadic-*E* is due to strong
shear winds which often develop near the maximum of the *E*-layer.

The Spread-F Ionograms occasionally show a large spread in the equivalent
height from which the *F*-region echoes are returned. This time spread,
which is much broader than the time width of the transmitted radio pulses,
is produced either by a blobby structure of the *F*-region which causes in
depth multiple scattering, or by a wavy structure of the *F*-region which
permits the reflection of the radio waves by curved surfaces at different
distances from the vertical. This phenomenon might last sometimes for
several hours and is usually a good indication of disturbed conditions in
the ionosphere.

The Ionospheric Irregularities, which represent local perturbations by a
few per cent in the electron density of the ionosphere. These irregularities
are often elongated along the lines of the earth's magnetic field and their
dimensions are of the order of 1 to 10 km. The ionospheric irregularities
usually move over the ground with velocities of the order of 100 to 200 m/s.

Their motion is predominantly in the $E - W$ direction, though it seems
to change during the course of the day and to have a seasonal dependence.
These irregularities are responsible for the *ionospheric scintillations*, i.e.,
of amplitude and phase fluctuations imposed on radio waves passing
through the ionosphere. By pointing our antennas toward a natural radio
source (e.g. Cassiopeia A) or toward a transmitting satellite, it is often
possible to recognize common scintillation features in the records obtained
by stations separated by a few kilometers. These common patterns usually
appear with certain time delays in the scintillation records of the two stations
because, as seen from Figure 2.6-I, the irregularities drift over the ground
with a velocity V. In a sense it is similar to the motion over the ground
of the shadow pattern from a cloud formation which is pushed by the wind.

When we have scintillation records from three different stations, it is
in general possible to compute the speed and direction of the motion of
the ionospheric irregularities. Let V be the *drift velocity* with which the

pattern of the irregularities is moving over the ground and let the vector V make an angle α with the line from station C to station A and an angle $(\theta - \alpha)$ with the line from station C to station B. The *apparent velocity* V'_{AC} along the AC line is,

$$V'_{AC} = \frac{(AC)}{t_A - t_C} \qquad (2.6\text{-}1)$$

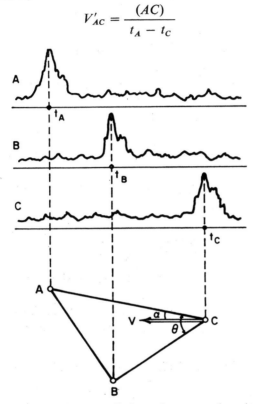

FIGURE 2.6-I   Simultaneous records from three ground stations, a few kilometers apart, showing the passage of an ionospheric irregularity over these stations

In practice we obtain first an average time delay either by averaging many individual measurements or better by doing a *cross-correlation*. In this method we multiply point by point the scintillation records of the two stations and we repeat this process several times shifting each time one of the two records by a certain time interval. The time shift that maximizes the product gives the best value for the average time delay between the two stations.

If we can assume that on the average we have an isometric pattern of irregularities (something like round clouds of different sizes) or that the irregularities are elongated (cigar-like) but on the average with their long axis at right angles to the direction of motion, then we can find V and α

without great difficulty. As seen from Figure 2.6-I, in the time interval $t_A - t_C$ the irregularity pattern will have advanced a distance equal to $(AC) \cos \alpha$. Therefore the drift velocity V is given by the relation,

$$V = \frac{(AC) \cos \alpha}{t_A - t_C} = V'_{AC} \cos \alpha \qquad (2.6\text{-}2)$$

and in the same manner we also have that,

$$V = \frac{(BC) \cos (\theta - \alpha)}{t_B - t_C} = V'_{BC} \cos (\theta - \alpha) \qquad (2.6\text{-}3)$$

From (2.6-2) and (2.6-3) we obtain,

$$\frac{V'_{AC}}{V'_{BC}} = \frac{\cos (\theta - \alpha)}{\cos \alpha} = \cos \theta + \sin \theta \tan \alpha \qquad (2.6\text{-}4)$$

which gives the angle α, i.e. the direction of motion of the irregularities, in terms of known parameters,

$$\tan \alpha = \frac{1}{\sin \theta} \left(\frac{V'_{AC}}{V'_{BC}} - \cos \theta \right) \qquad (2.6\text{-}5)$$

From (2.6-5) we can also compute the expression for $\cos \alpha$,

$$\frac{1}{\cos \alpha} = (1 + \tan^2 \alpha)^{1/2} = \frac{1}{\sin \theta} \left\{ \left(\frac{V'_{AC}}{V'_{BC}} \right)^2 - 2 \left(\frac{V'_{AC}}{V'_{BC}} \right) \cos \theta + 1 \right\}^{1/2} \qquad (2.6\text{-}6)$$

and finally from (2.6-2) and (2.6-6) we can obtain the expression for V,

$$V = \frac{V'_{AC} V'_{BC} \sin \theta}{[(V'_{AC})^2 + (V'_{BC})^2 - 2V'_{AC} V'_{BC} \cos \theta]^{1/2}} \qquad (2.6\text{-}7)$$

Ionospheric irregularities have been studied extensively since the end of the second world war by many groups around the world. The analysis given above represents only the first order solution and naturally one can find in the literature more sophisticated methods (Phillips and Spencer, 1955) which use auto-correlation and cross-correlation functions to take into account anisometric patterns and random components of the drift velocity.

Travelling Ionospheric Disturbances   These are large size perturbations of the electron density extending sometimes over 1000 km. They have been observed to travel with speeds of the order of 300 m/s over large distances and occasionally to make a full circle around the globe. The mechanism causing these large scale disturbances is not well understood. One possible suggestion is that they are produced by the sudden precipitation of a large number of energetic particles either in the polar regions or in the vicinity of a magnetic anomaly.

The Mid-Latitude Trough   This is a minimum in the electron densities of the ionosphere which develops primarily during the nighttime at a geomagnetic latitude (dip latitude) of approximately 60 degrees. Figure 2.6-II shows clearly the structure of the mid-latitude trough in the upper ionosphere as it was obtained through top side sounding with the Alouette I satellite (Chan and Colin, 1969). The mid-latitude trough, which was discovered only in recent years, is believed to occur around the latitude where the geomagnetic shell of the plasmapause crosses the horizontal layers of the ionosphere. One possible explanation is that the upward diffusion of the ionospheric plasma is inhibited inside the domain of the plasmasphere because the tubes of force in this region are already filled up with plasma. Outside the plasmapause, on the other hand, the tubes of force are essentially open-ended and the electron density of the ionosphere decreases substantially by the practically unrestricted upward diffusion of its ionization. This is sometimes called the *polar wind.*

According to this theory one would have expected a step function rather than a trough, but a trough could develop if right after the minimum we had a natural increase of the ionospheric electron density in the poleward direction. This actually is quite possible because at high latitudes the nearly tangential rays of the sun become very ineffective and they are replaced

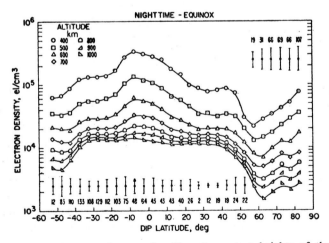

FIGURE 2.6-II   Average electron densities at constant heights of the upper ionosphere, which clearly show the mid-latitude through (Chan and Colin, 1969)

by low energy cosmic rays as the principal source of ionization. Due to the dipole configuration of the earth's magnetic field, the cosmic ray flux and therefore the electron density of the ionosphere increases continuously toward the pole because cosmic rays are charged particles and prefer to travel along, rather than across, magnetic field lines.

An alternative explanation, or possibly a contributing factor for the formation of the mid-latitude trough, is that due to the turbulance which must occur in the geomagnetic shells of the plasmapause, we have always a larger dumping of energetic particles along these field lines. The energy deposited by these particles into the ionosphere increases the temperature and therefore the recombination rate. Naturally a higher loss rate produces a lower electron content in this region of the ionosphere.

The geomagnetic latitude of the trough is a function of local time and tends to drift toward lower latitudes during geomagnetic disturbances. This is not unexpected because, as we have seen, during geomagnetic storms the plasmasphere decreases in size and the plasmapause moves closer to the earth at the equatorial plane. As a result the field lines of the plasmapause will cross the ionospheric layers at lower latitudes and the trough will move toward the equator. The study of the ionospheric trough and the plasmapause during quiet and disturbed conditions has attracted a great deal of interest in the last few years because it ties together in a very real but also very intricate manner the solar wind, the magnetosphere and the ionosphere.

Sudden Ionospheric Disturbances (SID)   These are caused by the enhanced ultraviolet and x-ray radiation from the sun during solar flare events. They occur only in the sun-lit side of the earth and they last, like the solar flares, from a few minutes to about one hour. We will discuss these disturbances in greater detail in Chapter 6.

Ionospheric Storms   These are closely associated with geomagnetic storms and can last from one to four days affecting the ionosphere over the entire globe. Observations of the unusual behavior of the ionosphere during these storms have been made and continue to be made by many groups around the world. Many diurnal, seasonal, and latitudinal storm effects have been discovered and serious efforts have been made for their theoretical interpretation. As we will see in Chapter 6, the theoretical problem is quite formidable because there are many factors that can be responsible for these effects. Furthermore, one must take into consideration the coupling of the ionosphere with the plasmasphere and the magnetosphere both of which are also under disturbed conditions during these storms. As a result there is still plenty of work to be done in this area, but the study of the ionosphere during disturbed conditions will undoubtedly help us understand also the complexities of the quiet ionosphere because several of the storm effects are exaggerated cases of some of the ionospheric anomalies we discussed in the beginning of this section.

2.7 Bibliography

A. Books for Further Studies

1. *The Magneto-Ionic Theory and its Applications to the Ionosphere*, J. A. Ratcliffe, Cambridge University Press, Cambridge, Great Britain, 1959.
2. *Radio Ray Propagation in the Ionosphere*, J. M. Kelso, McGraw-Hill, New York, 1968.
3. *Ionospheric Radio Propagation*, K. Davies, National Bureau of Standards Monograph 80, U. S. Printing Office, Washington, D. C., 1965.
4. *Whistlers and Related Ionospheric Phenomena*, R. A. Helliwell, Stanford University Press, Stanford, Calif., 1965.
5. *Introduction to Ionospheric Physics*, Henry Rishbeth and Owen K. Garriott, Academic Press, New York, N. Y., 1969.
6. *Physics of the Lower Ionosphere*, R. C. Whitten and I. C. Poppoff, Prentice Hall, Englewood Cliffs, N. J., 1965.
7. *Physics of the Upper Atmosphere*, ed. by J. A. Ratcliffe, Academic Press, N. Y., 1960.
8. *Radio Wave Absorption in the Ionosphere*, ed. by N. C. Gerson, Pergamon Press, New York, 1962.
9. *Radio Astronomical and Satellite Studies of the Atmosphere*, ed. by J. Aarons, North-Holland Publ. Co., Amsterdam, 1963.
10. *Satellite Environment Handbook*, ed. by F. S. Johnson, Stanford University Press, Stanford, Calif., 1965.
11. *Geophysics the Earth's Environment*, ed. by C. DeWitt, J. Hieblot and A. Lebeau, Gordon and Breach, New York, 1962.
12. *Space Physics*, ed. by D. P. LeGalley and A. Rosen, John Wiley and Sons, New York, 1964.
13. *Research in Geophysics*, vol. I, ed. by H. Odishaw, M. I. T. Press, Cambridge, Mass. 1964.
14. *Physics of the Earth's Upper Atmosphere*, ed. by C. O. Hines, I. Paghis, T. R. Hartz and J. A. Fejer, Prentice Hall, Englewood Cliffs, N. J., 1965.
15. *Introduction to Space Science*, ed. by W. N. Hess and G. D. Mead, Gordon and Breach, New York, 1968.

B. Articles in Scientific Journals

Allen, C. W., The interpretation of the XUV solar spectrum, *Space Sci. Rev.*, **1**, 91, 1965.
Brice, N. M., Bulk motion of the magnetosphere, *J. Geophys. Res.*, **72**, 5193, 1967.
Carlson, H. C., Ionospheric heating by magnetic conjugate-point photo-electrons, *J. Geophys. Res.*, **71**, 195, 1966.
Carpenter, D. L., Whistler studies of the plasmapause in the magnetosphere, *J. Geophys. Res.*, **71**, 643, 1966.
Chan, K. L. and L. Colin, Global electron density distribution from topside soundings. *Proc. IEEE*, **57**, 990, 1969.

Chapman, S., The production of ionization of monochromatic radiation incident upon a rotating atmosphere, *Proc. Phys. Soc.*, **43**, 26 (part I), 483 (part II), 1931 a, b.

Mendillo, M. and M. D. Papagiannis, An estimate of the dependance of the magnetospheric electric field on the velocity of the solar wind, *J. Geophys. Res.*, **76**, 6939, 1971.

Nicolet, M., The collision frequency of electrons in the ionosphere, *J. Atm. Terr. Phys.*, **3**, 200, 1953.

Papagiannis, M. D. and H. Mullaney, The geographic distribution of the ionospheric evening anomaly and its relation to the global pattern of neutral winds, *J. Atm. Terr. Physics*, **33**, 451, 1971.

Philips, G. J. and M. Spencer, The effects of anisometric amplitude patterns in the measurement of ionospheric drifts, *Proc. Roy. Soc.*, **68**, 481, 1955.

Ratcliffe, J. A., E. R. Schmerling, C. S. Setty and T. O. Thomas, The rates of production and loss of electrons in the *F*-region of the ionosphere, *Phil. Trans. Roy. Soc.*, A **284**, 621, 1956.

Sawyer, C., Statistics of solar active regions, *Ann. Rev. Astron. Astrophys.*, **6**, 115, 1968.

CHAPTER 3

THE MAGNETOSPHERE

3.1 The Earth's Magnetic Field

Present theories believe that the earth's magnetic field arises from electric currents flowing in the molten metallic core of the planet, which has a radius approximately one-half the radius of the earth. The currents are attributed to a dynamo mechanism operating inside this core. Recent discoveries suggest that the strength and orientation of the terrestrial magnetic field have changed considerably over geological periods. There is also strong evidence that the earth's magnetic field has reversed its direction several times during the lifetime of our planet.

Deep-space probes have discovered that both Mars and Venus lack a magnetic field. Jupiter, on the other hand, has a strong magnetic field the presence of which is inferred from the intense synchrotron radiation emitted by the giant planet. Venus lacks a magnetic field most probably because it rotates at an extremely slow rate around its axis. Though Venus must possess a molten core, since its size is very similar to the size of the earth, its very slow rotation probably prevents the formation of electric currents in the molten core. Mars, on the other hand, which rotates about its axis with a period very similar to that of the earth, is probably too small to have a molten core and therefore a magnetic field. Thus, molten core and rapid rotation are the prerequesites for the development of a planetary magnetic field. The earth and Jupiter satisfy both of these requirements, whereas Mars and Venus satisfy only one and for this reason they do not have a magnetic field.

The earth's magnetic field resembles a dipole field. Large scale regional departures from the dipole field are called *geomagnetic anomalies* and are attributed to irregularities or eddies in the dynamo current system. There are also smaller size anomalies due to local mineral deposits which are called *surface magnetic anomalies* and are helpful in locating these deposits of ferromagnetic materials. A very small component (less than 0.1%) of the terrestrial magnetic field is due to currents of charged particles in the outer atmosphere of the earth. The *Sq-current*, e.g., which circulates between the sunlit and the dark hemisphere, is responsible for the small, regular diurnal variations of the magnetic field observed on the ground. After a large solar flare, however, the enhanced flux of energetic particles from the sun can produce strong *ring currents* of charged particles around the earth that can produce large fluctuations (occasionally as high as 3%) of the terrestrial magnetic field. These magnetic disturbances, which might last for several days, are described by the different indices of geomagnetic activity.

The index K is the one most commonly used, and it is determined by the different observatories for consecutive three hour periods starting from 00:00 *Universal Time* (UT or Greenwich Time). The K index, in a scale from 0 to 9, describes the difference between the highest and the lowest deviation from the normal level of the magnetic field in any particular 3 hour interval. The value 9 represents a deviation in excess of 500 γ ($1 \gamma = 10^{-5}$ Gauss) for a mid-latitude station. The index Kp is called the *planetary 3-hour index* and is computed by averaging the K indices from a selected number of stations around the globe. It also follows a 0 to 9 scale but it is subdivided into 28 steps with the introduction of the + and − signs (0_0, $0+$, $1-$, 1_0, $1+$, ..., $8-$, 8_0, $8+$, $9-$, 9_0). The Kp index gives a measure of the geomagnetic activity over the entire globe during a 3-hour period. The sum of all eight Kp indices during a 24-hour period, which can range from 0 to 72, is called the ΣKp index and represents the global activity during a particular day.

An even better index for this purpose is the *daily equivalent planetary amplitude Ap*, which uses a linear scale (in essence the deviations in γ's) rather than the semi-logarithmic scale of the K index, thus giving a more realistic picture of the mean level of geomagnetic activity during a 24-hour period. The *daily character figure C*, which rates the magnetogram of a station during a 24-hour Greenwich day as very quiet ($C = 0$), moderately disturbed ($C = 1$) or severely disturbed ($C = 2$), is one more of the many other indices that are used by the experts to describe the level of geomagnetic activity. The importance of all these indices is that they allow an easy correlation of other phenomena, such as the different ionospheric variations, with the geomagnetic activity which in turn is a readily available indicator of the solar corpuscular flux and of the different electric currents which are generated during disturbed periods in the ionosphere and the magnetosphere of the earth.

The magnetic field of the earth can be represented, to a good approximation, by a dipole field with a magnetic moment $M = 8.05 \pm 0.02 \times 10^{25}$ Gauss cm³. The intensity of the field at the equator is ∼0.3 Gauss and at the poles ∼0.6 Gauss. The main field has a secular change of about 0.1% per year, and for this reason geomagnetic maps usually specify the epoch (year) for which they were constructed. It should be noted that in the CGS system the unit for the magnetic intensity H is the *Oersted*. Many people, however, measure the planetary and interplanetary magnetic fields in *Gauss* which is the unit of the magnetic induction B. The reason is that for non-magnetic media, such as the atmosphere, the magnetic permeability in the CGS system is equal to one and the two units are equivalent.

The centered dipole (its axis passes through the center of the earth) which fits best the earth's magnetic field has its axis directed along the line 79°N, 290°E to 79°S, 110°E. These are referred to respectively as the *north geomagnetic pole* and the *south geomagnetic pole*. The actual *magnetic poles* are asymmetrically located at 73°N, 262°E, and 68°S, 145°E (for epoch 1945), and for this reason the earth's magnetic field can be described exactly only through a long expansion in spherical harmonics. The dipole field is the first, and strongest term of this expansion which in several models that have appeared in the literature can have as many as 300 terms. Higher order terms (quadrapole, octapole, etc.) decrease faster with radial distance than the dipole field and, therefore, at distances of several earth radii the dipole field becomes an almost perfect approximation. For this reason, and because of its mathematic simplicity, the dipole approximation is used extensively in all kinds of geomagnetic problems.

The coordinates of the dipole field are called the *geomagnetic latitude* and the *geomagnetic longitude*. The angle the magnetic field makes with the horizontal is called the *inclination* or *dip* of the magnetic field, and the angle its horizontal component makes with the local geographic meridian is called the *declination* of the magnetic field.

3.2 The Dipole Magnetic Field

Let us consider the magnetic field at a point 0, a distance r_1 and r_2 from the two poles of a magnetic dipole. Let m be the strength of each of the two poles and d the distance that separates them. The product $M = md$ is called the *dipole magnetic moment*. The magnetic field at 0 will have the two components,

$$H_1 = \frac{m}{r_1^2} \quad \text{and} \quad H_2 = \frac{m}{r_2^2} \tag{3.2-1}$$

As seen from Figure 3.2-I, the distance r of the point 0 from the center of the dipole is given by the expressions,

$$r = r_1 \cos \alpha_1 + \frac{d}{2} \cos \theta$$

$$\tag{3.2-2}$$

$$r = r_2 \cos \alpha_2 - \frac{d}{2} \cos \theta$$

From the same diagram and the law of sines one can also see that,

$$\frac{d}{2}\sin\theta = r_1 \sin\alpha_1 = r_2 \sin\alpha_2 \qquad (3.2\text{-}3)$$

When $r_1 \simeq r_2 \gg d$, the angles α_1 and α_2 tend to zero and the respective cosines to unity. Setting $\cos\alpha_1 \simeq \cos\alpha_2 \simeq 1$ in (3.2-2), we obtain the relations,

$$(r_2^3 - r_1^3) \simeq \left(r + \frac{d}{2}\cos\theta\right)^3 - \left(r - \frac{d}{2}\cos\theta\right)^3 \simeq 3r^2\, d\cos\theta \qquad (3.2\text{-}4)$$

$$(r_1^3 + r_2^3) \simeq 2r^3 \qquad (3.2\text{-}5)$$

$$r_1^3 r_2^3 \simeq r^6 \qquad (3.2\text{-}6)$$

The radial component H_r of the magnetic field at a point 0 a large distance from the dipole is given by the expression,

$$H_r = H_1 \cos\alpha_1 - H_2 \cos\alpha_2 = \frac{m}{r_1^2}\left\{\frac{r - (d/2)\cos\theta}{r_1}\right\}$$

$$-\frac{m}{r_2^2}\left\{\frac{r + (d/2)\cos\theta}{r_2}\right\} = \frac{m\{r(r_2^3 - r_1^3) - (d/2)\,(r_2^3 + r_1^3)\cos\theta\}}{r_1^3 r_2^3}$$

$$= \frac{m\{r\,3r^2\,d\cos\theta - (d/2)\,2r^3\cos\theta\}}{r^6} = \frac{2m\,d\cos\theta}{r^3} = 2\frac{M}{r^3}\cos\theta$$

$$(3.2\text{-}7)$$

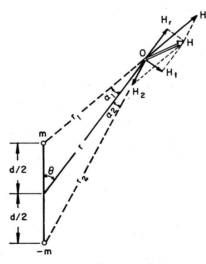

FIGURE 3.2-I   The intensity and direction of the magnetic field at a large distance from the dipole

and the tangential component H_t by the expression,

$$H_t = H_1 \sin \alpha_1 + H_2 \sin \alpha_2 = \frac{m}{r_1^2}\left\{\frac{(d/2)\sin\theta}{r_1}\right\} + \frac{m}{r_2^2}\left\{\frac{(d/2)\sin\theta}{r_2}\right\}$$

$$= \frac{m\,d\sin\theta}{2}\left(\frac{r_1^3 + r_2^3}{r_1^3 r_2^3}\right) = \frac{m\,d\sin\theta}{r^3} = \frac{M}{r^3}\sin\theta \qquad (3.2\text{-}8)$$

The line connecting the two poles of the dipole defines the axis of the north and south magnetic poles. Therefore, the angle θ, as seen from Figure 3.2-I, represents the *geomagnetic co-latitude*. From (3.2-7) and (3.2-8) it follows that at the poles, where $\theta = 0$, the magnetic field H_p is all in the radial direction and is given by the expression,

$$H_p = 2\frac{M}{r^3} \qquad (3.2\text{-}9)$$

while at the equator, where $\theta = 90°$, the magnetic field H_e is entirely in the tangential direction and is given by the expression,

$$H_e = \frac{M}{r^3} \qquad (3.2\text{-}10)$$

Thus the magnetic field at the poles has twice the intensity of the magnetic field at the equator, i.e. $H_p = 2H_e$.

We can now compute the intensity of the total magnetic field at any geomagnetic co-latitude θ from the radial (3.2-7) and the tangential (3.2-8) components we have already obtained,

$$H = (H_r^2 + H_t^2)^{1/2} = \left\{\left(2\frac{M}{r^3}\cos\theta\right)^2 + \left(\frac{M}{r^3}\sin\theta\right)^2\right\}^{1/2}$$

$$= \frac{M}{r^3}(1 + 3\cos^2\theta)^{1/2} \qquad (3.2\text{-}11)$$

Equation (3.2-11) shows that the intensity of the dipole magnetic field decreases with distance as the third power (r^3) of the radial distance and for the same r varies, as we have seen already, by a factor of 2 from the poles to the equator. Using (3.2-10) and (3.2-11) we can express the total magnetic field at any given value of θ in terms of the equatorial magnetic field at the same radial distance,

$$H = H_e(1 + 3\cos^2\theta)^{1/2} \qquad (3.2\text{-}12)$$

From the radial (vertical) and tangential (horizontal) components of the magnetic field, we can also find the inclination angle ω (dip) of the field,

$$\tan\omega = \frac{H_r}{H_t} = \frac{2H_e\cos\theta}{H_e\sin\theta} = 2\cot\theta \qquad (3.2\text{-}13)$$

5*

Equation (3.2-13) shows that the dip of the magnetic field is independent of the radial distance and therefore the magnetic field at any altitude above a given station will always be parallel to the magnetic field on the ground. The intensity of the field, however, will decrease as the cube of the radial distance and will be given by the expression,

$$H(h) = H_g \left(\frac{R_0}{R_0 + h} \right)^3 \tag{3.2-14}$$

An imaginary line to which the magnetic field is always tangential is called a *line of force* or a *field line*. For such a line, as seen from Figure 3.2-II, we have,

$$dr = (r\, d\theta) \tan \omega = r\, d\theta\, 2 \cot \theta \tag{3.2-15}$$

Integrating (3.2-15) we get,

$$r = r_0 \sin^2 \theta \tag{3.2-16}$$

where r_0 is the radial distance of the field line at the equator. Equation (3.2-16) is a very important relation because it gives the geometry of the field lines of the dipole field. The magnetic field at the equatorial crossing of a given field line is usually denoted by H_0 and it is a very important parameter because by knowing H_0 we can determine the magnetic field at any point along this field line. By combining (3.2-11), (3.2-16), and (3.2-10) we obtain,

$$H = \frac{M}{r^3} (1 + 3 \cos^2 \theta)^{1/2} = \frac{M}{r_0^3} \frac{(1 + 3 \cos^2 \theta)^{1/2}}{\sin^6 \theta}$$

$$= H_0 \frac{(1 + 3 \cos^2 \theta)^{1/2}}{\sin^6 \theta} \tag{3.2-17}$$

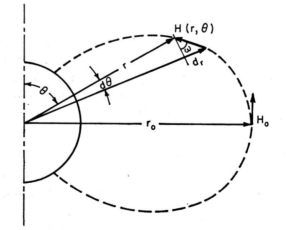

FIGURE 3.2-II   The geometry of the field line of a dipole magnetic field

Equation (3.2-17) is probably the most useful expression in the mathematical description of the dipole magnetic field.

Two more parameters of the dipole field which are often used in practical applications are the *arc length (ds)* along a field line, and the *radius of curvature (ϱ)* of the field line at any point along its curved path. Using the corresponding expressions of analytic geometry we find,

$$ds = (\dot{x}^2 + \dot{y}^2)^{1/2} \, d\theta = r_0(1 + 3\cos^2\theta)^{1/2} \sin\theta \, d\theta \qquad (3.2\text{-}18)$$

$$\varrho = \frac{1}{K} = \frac{(\dot{x}^2 + \dot{y}^2)^{3/2}}{(\ddot{x}\dot{y} - \dot{y}\ddot{x})} = \frac{r_0}{3} \frac{(1 + 3\cos^2\theta)^{3/2}}{(1 + \cos^2\theta)} \sin\theta \qquad (3.2\text{-}19)$$

In (3.2-18) and (3.2-19) x and y have the values,

$$x = r\sin\theta = r_0 \sin^3\theta, \quad y = r\cos\theta = r_0 \sin^2\theta\cos\theta \quad (3.2\text{-}20)$$

and the dots represent differentiation with respect to θ which is the common parameter.

A very useful concept is to represent the intensity of the magnetic field by the *flux density* of the field lines. In this way we can say that 1 Gauss is equal to a certain number of field lines crossing a unit area normal to the lines. The total number of lines crossing a given surface is called the *total magnetic flux* through this surface, and in more mathematical terms it is the integral of the flux density over the entire surface. A group of field lines forms a *tube of force* which might converge or diverge but contains always the same number of lines. Consequently the total magnetic flux through any cross section of a tube of force remains constant.

To become more familiar with this very useful concept, let us compute the total magnetic flux crossing one of the two polar caps of the earth. The area on the surface of the earth between co-latitudes θ and $\theta + d\theta$ is,

$$dA = 2\pi(R_0 \sin\theta)(R_0 \, d\theta) = 2\pi R_0^2 \sin\theta \, d\theta \qquad (3.2\text{-}21)$$

where R_0 is the radius of the earth.

Multiplying dA times the component of the magnetic field that is normal to the surface dA (in this case the radial component H_r), we find that the total flux through the area dA is,

$$dF = H_r \, dA = \left(2\frac{M}{R_0^3}\cos\theta\right)(2\pi R_0^2 \sin\theta \, d\theta) \qquad (3.2\text{-}22)$$

and by integrating (3.2-22) from $\theta = 0$ to the edge of the polar cap $\theta = \theta_0$, we obtain the total magnetic flux F through the polar cap.

$$F = 2\frac{M}{R_0^3} 2\pi R_0^2 \left(\frac{\sin^2\theta_0}{2}\right) = \left(2\frac{M}{R_0^3}\right)\{\pi(R_0 \sin\theta_0)^2\} = H_{gp} A_0 \qquad (3.2\text{-}23)$$

Equation (3.2-23) gives the interesting result that the total magnetic flux through the polar cap is the same as if a uniform magnetic field equal and parallel to the ground field at the pole (H_{gp}) were crossing an area A_0 equal to the flat base of the polar cap.

Because the total magnetic flux is conserved, the result of (3.2-23) must be equal to the total magnetic flux (the same total number of field lines) crossing the equatorial plane beyond the radial distance r_0 which corresponds to the field lines starting at the co-latitude θ_0. Making use of the geometry of the field line (3.2-16), we can easily find r_0 in terms of R_0 and θ_0,

$$r_0 = \frac{R_0}{\sin^2 \theta_0} \tag{3.2-24}$$

As seen from Figure 3.2-III, the magnetic flux crossing the polar cap between θ and $\theta + d$ will cross the equatorial plane between the radial distances r and $r + dr$. The magnetic flux through this ring in the equatorial plane, to which H_e is normal, is,

$$dF = H_e \, dA = \left(\frac{M}{r^3}\right)(2\pi r \, dr) \tag{3.2-25}$$

Integrating (3.2-25) from $r = r_0$ to infinity and making use of (3.2-24), we find,

$$F = \frac{2\pi M}{r_0} = \frac{2\pi M}{R_0}\sin^2\theta_0 = \left(\frac{2M}{R_0^3}\right)\{\pi(R_0 \sin \theta_0)^2\} \tag{3.2-26}$$

As expected the result of (3.2-26) confirms the result of (3.2-23).

The conservation of the total magnetic flux in a tube of force, or for that matter in a shell of force enclosing the earth as we did here, is used in many types of computations related to the magnetic field. One could

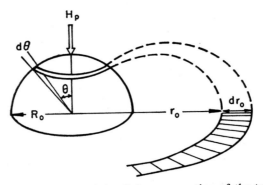

FIGURE 3.2-III   The principle of the conservation of the total magnetic flux in a tube of force

have derived, for example, the relation $R_0 = r_0 \sin^2 \theta_0$ of (3.2-24), by simply equating (3.2-23) to $2\pi M/r_0$ which was the corresponding total magnetic flux at the equatorial plane. We will have an opportunity to see another application of this method when we will compute in Section 3.5 the magnetic field in the tail of the magnetosphere.

3.3   Motions of Charged Particles in a Dipole Magnetic Field

A charged particle moving with velocity V at an angle ψ, called the *pitch angle*, to a magnetic field will experience the Lorentz force,

$$F = \frac{e}{c} V \times H = \frac{e}{c} VH \sin \psi \qquad (3.3\text{-}1)$$

which will set the particle in a helical motion around a line of force of the magnetic field. The Lorentz force is balanced by the centrifugal force produced by the component $V_n = V \sin \psi$ of the particle's velocity which is normal to the magnetic field, i.e.,

$$\frac{e}{c} VH \sin \psi = \frac{mV_n^2}{R_H} = \frac{mV^2 \sin^2 \psi}{R_H} \qquad (3.3\text{-}2)$$

where R_H is the radius of gyration around the field line which is called the *gyro-radius* or the *cyclotron radius*,

$$R_H = \left(\frac{mc}{eH}\right) V_n = \frac{mcV \sin \psi}{eH} \qquad (3.3\text{-}3)$$

Equation (3.3-3) defines also the *angular cyclotron frequency* ω_H which is given by the relation,

$$\omega_H = \frac{V_n}{R_H} = \frac{eH}{mc} \qquad (3.3\text{-}4)$$

from which we obtain also the expression for the *gyro-frequency* or *cyclotron frequency* f_H.

$$f_H = \frac{\omega_H}{2\pi} = \frac{eH}{2\pi mc} = 2.8H \qquad (3.3\text{-}5)$$

In the numerical form of (3.3-5) f_H is in MHz and H in Gauss. For relativistic velocities the mass m in all the above formulas is related to the rest mass m_0 of the particle by the well known expression of the special theory of relativity,

$$m = \frac{m_0}{\left(1 - \dfrac{V^2}{c^2}\right)^{1/2}} \qquad (3.3\text{-}6)$$

A charged particle moving in a circular orbit produces a magnetic moment μ equal to,

$$\mu = \frac{iA}{c} \qquad (3.3\text{-}7)$$

where i is the current, A the area of the loop $(= \pi R_H^2)$, and c the speed of light. The charged particle passes by a given point on its circular orbit $V_n/2\pi R_H$ times per second, and therefore the current, i.e., the amount of charge passing by a given point per second will be $i = eV_n/2\pi R_H$. Thus the magnetic moment μ of a charged particle gyrating in a magnetic field is given by the expression,

$$\mu = \left(\frac{eV_n}{2\pi R_H}\right)\left(\frac{\pi R_H^2}{c}\right) = \frac{e}{2c} V \sin \psi \, R_H \qquad (3.3\text{-}8)$$

When the cyclotron radius R_H is much smaller than the radius of curvature ϱ (3.2-19) of the field lines, the magnetic moment of the spiraling particle is an adiabatic invariant (constant) of the motion. The kinetic energy of the particle $E_k = \frac{1}{2} m V^2$ is also conserved, and therefore μ/E_k must be a constant of the motion of the particle, i.e.,

$$\frac{\mu}{E_k} = \frac{eV \sin \psi \, R_H/2c}{mV^2/2} = \frac{e \sin \psi R_H}{mVc} = \text{constant} \qquad (3.3\text{-}9)$$

Finally using (3.3-3) in (3.3-9) we find that,

$$\frac{\mu}{E_k} = \frac{\sin^2 \psi}{H} = \text{constant} \qquad (3.3\text{-}10)$$

This equation is the basic relation governing the motion of charged particles in a magnetic field. Since, as seen from (3.2-17), H changes along a field line and $\sin^2 \psi/H$ must remain constant it follows that ψ must change as the particle spirals along a line of force of the earth's magnetic field.

Let ψ_0 be the pitch angle that the particle's velocity makes with the magnetic field at the equator. From (3.3-10) and (3.2-17) we get,

$$\sin^2 \psi = \frac{H}{H_0} \sin^2 \psi_0 = \frac{(1 + 3 \cos^2 \theta)^{1/2}}{\sin^6 \theta} \sin^2 \psi_0 \qquad (3.3\text{-}11)$$

Thus the components of the particle's velocity that are normal V_n and parallel V_p to the magnetic field as the particle spirals along the field line are given by the expressions,

$$V_n^2 = V^2 \sin^2 \psi = V^2 \frac{H}{H_0} \sin^2 \psi_0 = V^2 \frac{(1 + 3 \cos^2 \theta)^{1/2}}{\sin^6 \theta} \sin^2 \psi_0$$

$$(3.3\text{-}12)$$

and,

$$V_p^2 = V^2 \cos^2 \psi = V^2 \left(1 - \frac{H}{H_0} \sin^2 \psi_0\right)$$

$$= V^2 \left\{1 - \frac{(1 + 3 \cos^2 \theta)^{1/2}}{\sin^6 \theta} \sin^2 \psi_0\right\} \qquad (3.3\text{-}13)$$

Equation (3.3-13) shows that the velocity along the line of force V_p will become zero at the point where $H = H_0/\sin^2 \psi_0$. At this point the particle is reflected and V_p, going through zero, reverses its direction. The co-latitude θ_m of the mirroring (turning) points for particles of given pitch angle ψ_0 is given, as seen from (3.3-13), by the transcendental equation,

$$\frac{\sin^6 \theta_m}{(1 + 3 \cos^2 \theta_m)^{1/2}} = \sin^2 \psi_0 \qquad (3.3\text{-}14)$$

Occasionally the equatorial pitch angle ψ_0 is so small that the magnetic field H of a field line cannot reach in time the large value $H_0/\sin^2 \psi_0$, and the particle is precipitated (lost) in the lower atmosphere. Usually, however, the equatorial pitch angle is large enough to permit the mirroring (reflection) of the particles above ~ 300 km, so that the particles can avoid the denser regions of the atmosphere where they are lost through collisions. These particles can remain trapped in the earth's magnetic field bouncing continuously between the northern and southern hemispheres and are often referred to as the *trapped radiation*. This is a somewhat confusing term, because the word "radiation" usually refers to a flux of photons. Beams of charged particles, however, are still called rays (cosmic rays) and in the early days of space research they wanted to distinguish the trapped protons and electrons (trapped radiation) from the free high-energy particles (cosmic radiation) which would also register in their counters.

The period of a mirror oscillation in a dipole field consists of four sym-metric (equal) parts and is given by the integral,

$$T_m = 4 \int_{\theta_m}^{\pi/2} \frac{ds}{V_p(\theta)} = \frac{4r_0}{V} \int_{\theta_m}^{\pi/2} \frac{\sin \theta (1 + 3 \cos^2 \theta)^{1/2} \, d\theta}{\left\{1 - \dfrac{(1 + 3 \cos^2 \theta)^{1/2}}{\sin^6 \theta} \sin^2 \psi_0\right\}^{1/2}} \qquad (3.3\text{-}15)$$

In (3.3-15) we have substituted ds by (3.2-18), and V_p by (3.3-13). Numerical solutions of (3.3-15) show that the period T_m, which typically is of the order of one second, can be approximated by the following semiempirical relation,

$$T_m = \frac{4r_0}{V} (1.30 - 0.56 \sin \psi_0) \qquad (3.3\text{-}16)$$

In addition to mirroring between the two conjugate hemispheres, the trapped particles have also an azimuthal motion with electrons drifting toward the east and protons toward the west. This phenomenon has actually two causes: The one is the centrifugal force acting on the particles as they move along the curved field lines. This force is proportional to V_p^2/ϱ and induces an azimuthal motion to the particles because it is like trying to move these particles outwards against a normal magnetic field. The other cause is that at larger radial distances the earth's magnetic field becomes weaker and therefore the cyclotron radius (3.3-3) becomes larger. As seen from Figure 3.3-I, this makes the charged particles drift sideways in a direction perpendicular to the gradient of the magnetic field. This effect is proportional to the normal component of the velocity V_n and the cyclotron radius R_H. The total drift velocity V_d due to the two causes we have mentioned is given (Spitzer, 1956) by the expression,

$$V_d = \frac{1}{\varrho \omega_H} \left(V_p^2 + \frac{1}{2} V_n^2 \right) \tag{3.3-17}$$

As the particles move a distance $ds = V_p dt$ along a field line, they also drift over an azimuthal angle $d\phi$ which, because the particles move at a co-latitude θ, is related to the drift velocity V_d by the relation.

$$(r \sin \theta) \, d\phi = V_d \, dt = V_d \left(\frac{ds}{V_p} \right) \tag{3.3-18}$$

Hence the azimuthal drift angle ϕ_m covered during one mirror oscillation is,

$$\phi_m = 4 \int_{\theta_m}^{\pi/2} \frac{V_d \, ds}{V_p r \sin \theta} = 4 \int_{\theta_m}^{\pi/2} \frac{(V_p^2 + \frac{1}{2} V_n^2) \, ds}{\varrho \omega_H V_p r \sin \theta} \tag{3.3-19}$$

Dividing ϕ_m by the period (3.3-15) of a mirror oscillation we obtain the *angular drift velocity* $\omega_d = \phi_m/T_m$. Numerical integrations of (3.3-19) have produced values for ϕ_m which together with the numerical results of (3.3-15) have allowed Hamlin *et al.* (1961), to obtain a simple semiempirical relation for the angular drift velocity ω_d,

$$\omega_d = \frac{3mcV^2}{eH_0 r_0^2} (0.35 + 0.15 \sin \psi_0) \tag{3.3-20}$$

Since $H_0 r_0^3$ is a constant equal to the magnetic moment M of the earth, and $\frac{1}{2} mV^2 = E_k$ is the kinetic energy of the particle, it follows that the period $T_d = 2\pi/\omega_d$ in which the particles make a complete circle of the earth is inversely proportional to $r_0 E_k$. For $\psi_0 = 90°$, the expression for T_d becomes

$$T_d = \frac{44}{LE_k} \text{(min)} \tag{3.3-21}$$

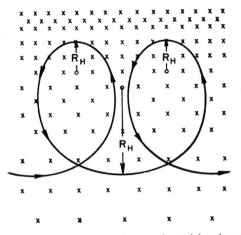

FIGURE 3.3-I   The azimuthal motion of trapped particles due to the decrease
of the earth's magnetic field with radial distance

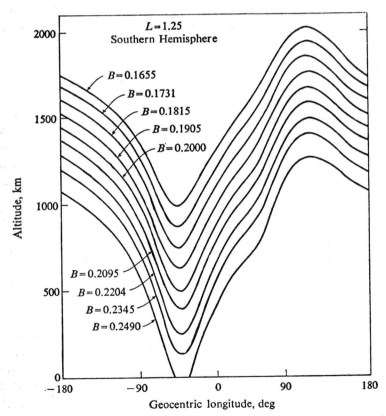

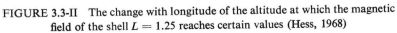

FIGURE 3.3-II   The change with longitude of the altitude at which the magnetic
field of the shell $L = 1.25$ reaches certain values (Hess, 1968)

where T_d is in minutes, L is the equatorial distance r_0 of the field line in earth radii, and E_k the kinetic energy of the particle in MeV. Thus typically T_d is of the order of one hour and particles usually perform many thousands of mirror oscillations while completing a full turn around the earth.

In a symmetric dipole magnetic field the mirror points at the two conjugate regions are completely equivalent. In the case of the earth, however, they are not exactly the same because the terrestrial magnetic field is not an ideal dipole. Furthermore there are also, as we have seen, large isolated regions on the earth where the magnetic field is weaker than the dipole field (geomagnetic anomalies). Thus, as the trapped particles drift around the globe, the altitude of the mirroring points varies as the intensity of the magnetic field changes. This is clearly seen in Figure 3.3-II. In the regions where the field becomes weaker, the particles will penetrate into lower altitudes and therefore in these areas we will observe the precipitation of a much larger number of trapped particles. The *South Atlantic anomaly* and the *Cape Town anomaly* are typical examples of several such regions that exist around the earth. The sharp dip around -30 degrees in Figure 3.3-II is due to the South Atlantic anomaly.

3.4   The Radiation Belts

The belts, or zones, of trapped radiation were the first major discovery of the space age. The first American satellite, Explorer I, was launched on January 31, 1958, carrying among other instruments a Geiger counter provided by Van Allen's group of the University of Iowa. Readings from this counter were obtained only when the satellite passed above a small number of tracking stations. When the satellite pass was a low one, the counting rate was the one expected from the known cosmic ray flux. When Explorer I was near apogee, however, the corresponding ground station received the message that the counting rate of the Geiger counter had dropped to zero.

This unexpected result meant either that the counter was malfunctioning or that there was no radiation at higher altitudes, which did not seem to make much sense. A third possibility was pointed out by Carl McIlwain, who suggested that the zero counting rate might be due to the saturation of the counter (dead-time effect) from a very high flux of energetic particles. McIlwain's suggestion was confirmed by Explorer III after Explorer II failed to reach an orbit. The conclusion from these results was that at higher altitudes the satellites enter the region of trapped radiation where the counting rate is 1,000 times higher than what it would have been due to the cosmic ray flux.

It is interesting to note parenthetically that the Russians had placed Geiger counters on Sputnik II, which was launched before Explorer I, and therefore could have discovered the radiation belts before the Americans. The Russian satellite, however, happened to be always underneath the radiation belts (near perigee) whenever the counter was monitored over the Soviet Union and the Russians missed a great opportunity. The discovery of the Iowa group was later confirmed by Sputnik III, but the honor of discovering the radiation zones belongs to Prof. Van Allen and his group, and for this reason the belts of trapped radiation are often called the Van Allen belts.

It should be mentioned that several people had already worked on the theory of trapped radiation way before the actual discovery of the radiation belts. Störmer, for example, had calculated in detail the orbits of particles trapped in a dipole magnetic field, and Singer (1957) had suggested the possible existence of zones of trapped radiation. Christophilos, on the other hand, was preparing at that time to test the ability of the earth's magnetic field to trap the high energy charged particles released by a nuclear explosion at high altitudes. This project was actually carried out in 1958 by the Department of Defense with the code name *Argus*. Under the direction of Christophilos (1959), three small (1 kt) nuclear bombs were detonated at high altitudes and created the first belt of artificial radiation. The results of these tests were studied extensively by Explorer IV. The injection of artificial radiation into the earth's magnetic field was repeated on a much larger scale in 1962, when a 1.4 Mt nuclear bomb was detonated 400 km above the Johnson Islands in the Pacific Ocean. This was called the *Starfish* or *Fishbowl* project and produced an intense and long-lasting artificial belt of trapped radiation. The Russians performed also several nuclear explosions in space in 1962, but such experiments are now banned by an international agreement on nuclear controls.

Following the first satellite observations, a great interest developed in exploring the morphology of the zones of trapped radiation. The first complete picture that emerged from the mapping of the belts is shown in Figure 3.4-I. This diagram depicts counting rates of particles energetic enough to penetrate the shielding of 1 gr/cm^2 of lead which covered the approximately 1 cm^2 window of the Geiger counter. As seen from Figure 3.4-I, the counting rates reached values higher than 10^4 counts/sec in two different regions. This led to the notion that there are actually two radiation belts which were named the *inner* and *outer* Van Allen belts.

The first counters could not differentiate between energetic protons and energetic electrons. Today, however, we know that the high counting rates of the inner belt are produced by energetic protons with energies in the

10 to 100 MeV range, while the high counting rates of the outer belt are produced by high energy electrons with energies in the 1 MeV range and above. This of course does not mean that there are no energetic electrons in the inner belt or energetic protons in the outer belt. Since the total trapped charge of either belt must be equal to zero it simply means that the electrons of the inner belt and the protons of the outer belt were not energetic enough to penetrate the lead foil that was shielding the window of the Geiger counter. It should be mentioned that the belts of trapped radiation also hold energetic alpha particles and even nuclei with $z \gg 3$ (Krimigis *et al.*, 1970). One should not forget also that the energetic particles of the belt move through a much higher number (nearly a million times more) of thermal ($E \sim 1$ eV) electrons and ions that constitute the plasmasphere of the earth.

More detailed studies have shown that the spatial distribution of the trapped electrons and protons varies with the energy of these particles. The highly energetic protons peak near $1.5R_0$, but for protons of lower energies the peak of the belt moves farther out and the width of the belt increases considerably. The same is also true for the energetic electrons

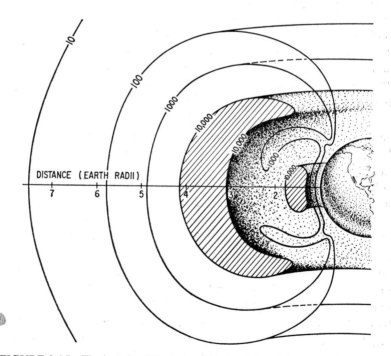

FIGURE 3.4-I   The inner and the outer radiation belts as they were first mapped by Van Allen's group of the University of Iowa (White, 1970)

of the outer belt. For energies above 1 MeV the belt is relatively narrow and peaks near $4R_0$ while for electrons of lower energies the belt spreads out fairly evenly over a much larger volume. These effects are demonstrated in Figure 3.4-II which shows the approximate distribution of trapped electrons and protons of lower energy compared to the ones shown in Figure 3.4-I. In general, as we examine the more plentiful particles of lower energies we find that the concept of an inner and an outer belt tends to disappear because these less energetic particles are distributed over a much wider range of L shells.

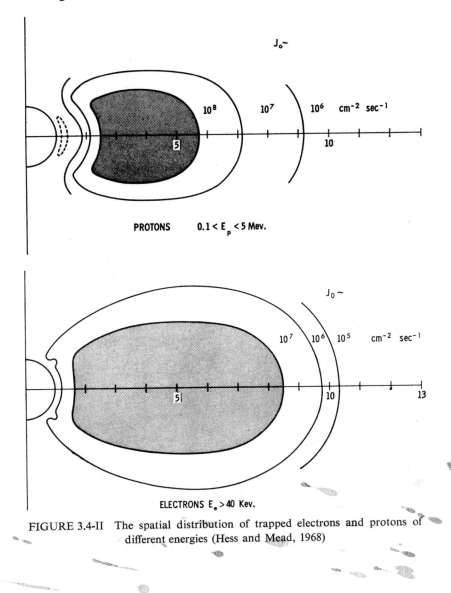

FIGURE 3.4-II The spatial distribution of trapped electrons and protons of different energies (Hess and Mead, 1968)

The expression *L-shell* comes from the *B and L coordinates* which were introduced by McIlwain (1961). In this system *B* is the intensity of the magnetic field and *L* is given by an integral of the motion of the trapped particles. In the dipole approximation, *L* turns out to be the equatorial distance r_0 of the field line on which these particles spiral. As seen from Figure 3.4-III, this is a very practical system because it allows the accurate plotting of the spatial distribution of the counting rates observed. Plotting the counting rates in *r* and *θ* coordinates gives a much more confused diagram because, as seen from Figure 3.3-II, for given *r* and *θ* the magnetic field *B* varies with longitude. For this reason diagrams like the ones of Figure 3.4-II can only be used as an average approximate mapping, while diagrams in the *B* and *L* system remain correct for any magnetic meridian.

The great usefulness of the *L* shell concept has been extended also in determining a meaningful magnetic latitude on the ground. From (3.2-16) we have seen that in the dipole field the geomagnetic latitude *λ* is given by the expression $\cos^2 \lambda = R_0/r_0$, where R_0 is the radius of the earth and r_0 the equatorial crossing of the field line which crosses this latitude. In an equivalent way, by replacing r_0/R_0 with the *L* shell value of the

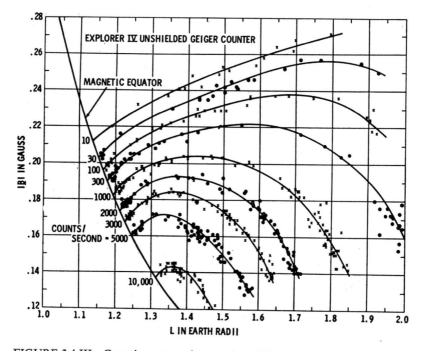

FIGURE 3.4-III Counting rates of trapped radiation plotted in the B and L coordinate system introduced by McIlwain (Hess and Mead, 1968)

McIlwain theory, we can obtain the *invariant latitude* Λ,

$$\cos^2 \Lambda = \frac{1}{L} \qquad (3.4\text{-}1)$$

which recently is used quite frequently as the most appropriate latitude for geomagnetic phenomena. Also, since many effects such as auroras, ionospheric storms, etc., occur at an altitude h above the ground a *generalized invariant latitude* Λ' has been introduced to account for the change of the L shell with radial distance.

$$\cos^2 \Lambda' = \left(1 + \frac{h}{R_0}\right)\bigg/ L \qquad (3.4\text{-}2)$$

Since the discovery of the Van Allen belts people have tried hard to find the source, or sources, of the energetic particles that become trapped in these zones. Originally it was thought that the source for the energetic protons of the inner belt was the *Cosmic Ray Albedo Neutron Decay* (CRAND). According to this mechanism, high energy cosmic ray protons colliding with atmospheric nuclei produce neutrons some of which move outward (albedo) and after a few minutes they decay into a proton, an electron, and an antineutrino (β-decay). These high-energy protons and electrons continuously populate the inner belt but are also continuously lost through inelastic collisions with atmospheric atoms and nuclear interactions. This is called the *leaky bucket* model because new particles are being added constantly while others leak out all the time. Thus the inner belt is like a leaky bucket in which water is poured in, while water leaks out, so as to maintain a steady level of water in the bucket.

More recent computations have shown that the CRAND process cannot provide the number of protons, especially those with $E < 30$ MeV, needed to explain the inner belt. Neutron decay of the solar proton albedo (SPAND) is also a contributing factor but does not appear to be a significant particle source. Several other suggestions have been made, such as neutral hydrogen particles carried by the solar wind, μ-mesons from cosmic rays, etc., but as of now the source of the inner belt particles is not well understood. The spatial distribution and particle fluxes of the inner belt remain nearly constant, changing only very slowly over long periods of time.

The outer belt peaks near $4R_0$, but high counting rates of low energy electrons are observed all the way up to the magnetopause (see next section). The basic characteristic of the outer belt is its variability. Changes of the counting rate by a factor of 10 have been observed to occur in intervals of a few days. The outer belt, contrary to the inner belt, is very sensitive to the solar activity and for this reason it is assumed that the solar wind is the particle source of the outer belt. This, however, is only an educated guess because we still do not understand how the solar particles get into the magnetosphere and populate the outer belt. We also do not know how the relatively slow particles of the solar wind are accelerated to the energies observed in the outer belt. People have proposed several acceleration mechanisms, such as strong electric fields and magnetic pinching effects. Also suggested were different possible pathways, such as through the dayside cusps (Figure 3.5-III) or the tail of the magnetosphere, by which the solar wind particles could enter into the geomagnetic covity. None of these theories has reached yet wide acceptance and though much has been learned about the earth's radiation belts, there are still many questions which continue to stimulate extremely interesting theoretical and experimental research programs in this area.

Space probes have found that Mars and Venus lack a magnetic field and therefore lack also any zones of trapped radiation. Jupiter, on the other hand, emits strong radio radiation which almost without any doubt is synchrotron radiation emitted by relativistic electrons trapped in the strong magnetic field of Jupiter. For a space observer the earth's Van Allen belts must also be a source of synchrotron radio emission with a maximum probably below 1 MHz. Synchrotron radiation is polarized in the plane of the orbit of the electron, i.e. in a plane normal to the magnetic field. The spectral distribution of the total power emitted by a single relativistic electron (Ginzburg and Svrovatskii, 1965) is given by the expression,

$$p(f) = \frac{3^{1/2} e^3 H \sin \psi}{m_0 c^2} \left(\frac{f}{f_c}\right) \int_{(f/f_c)}^{\infty} K_{5/3}(x) \, dx \qquad (3.4\text{-}3)$$

where $K_{5/3}(x)$ is the Bessel function of the second kind of order 5/3 and f_c is the *critical Schwinger frequency* given by the expression,

$$f_c = \frac{3eH \sin \psi}{4\pi m_0 c} \left(\frac{E_T}{m_0 c^2}\right)^2 \simeq 16 \times 10^{-6} \, H \sin \psi (E_T)^2 \qquad (3.4\text{-}4)$$

In (3.4-4) H is in Gauss, the total energy of the electrons E_T is in electronvolts, and f_c in Hz. As seen from Figure 3.4-IV, the synchrotron spectrum peaks at a frequency f_m directly related to f_c.

$$f_m \simeq 0.3 f_c \simeq \frac{1}{2} f_H \left(\frac{E_T}{m_0 c^2}\right)^2 \simeq 4.6 \times 10^{-6} \, H \sin \psi (E_T)^2 \qquad (3.4\text{-}5)$$

and the general shape of the spectrum can be approximated by,

$$p(f) \propto f^{1/3} e^{-f/f_c} \qquad (3.4\text{-}6)$$

The total power emitted over the entire spectrum is,

$$W = 1.57 \times 10^{-15} \, H^2 \sin^2 \psi \left(\frac{E_T}{m_0 c^2} \right)^2 \qquad (3.4\text{-}7)$$

where W is in erg/sec, H in Gauss, and E_T and $m_0 c^2$ are expressed in the same units.

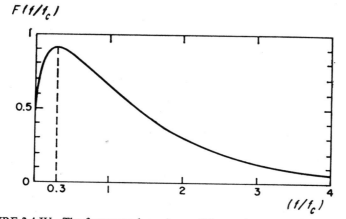

FIGURE 3.4-IV   The frequency dependance of the synchrotron radiation emitted by a relativistic particle

The *half-life* $\tau_{1/2}$ of the synchrotron emission is the period in which a charged particle emits half of its energy in the form of synchrotron radiation. This time interval is given by the expression,

$$\tau_{1/2} = \frac{E_T}{W} = \frac{2.6 \times 10^{14}}{E_T H^2 \sin^2 \psi} \, \sec \qquad (3.4\text{-}8)$$

where again E_T is given in eV, H in Gauss and $\tau_{1/2}$ in seconds. As we will see in Chapter 8, the half-life of synchrotron radiation is of great importance in many astrophysical and radioastronomical problems.

3.5   The Boundary and the Tail of the Magnetosphere

The magnetosphere is the region where the motion of the charged particles is primarily governed by the earth's magnetic field. Originally it was thought that the terrestrial magnetic field extends way out into the interplanetary space, becoming weaker with distance and gradually merging into the emptiness of free space. An alternative configuration for the earth's magnetic field was discussed for the first time in 1931 by Chapman and Ferraro. In their effort to understand the mechanism of the magnetic storms and

6*

their relation to the solar flares, Chapman and Ferraro (1931) suggested that a large flare is accompanied by the ejection from the sun of a big plasma cloud which reaches the earth in approximately one day. As this cloud blows past the earth, it exerts a pressure on the terrestrial magnetic field and sweeps it back in an aerodynamic configuration.

Under the pressure of this cloud, the earth's magnetic field is confined inside a region which is called the *magnetic cavity*. The boundary of this cavity is called the *magnetopause* and marks the end of the magnetosphere. At this boundary the pressure of the compressed tubes of force of the earth's magnetic field is equal and balances the pressure exerted by the stream of charged particles from the sun. The pressure is exerted almost exclusively by the protons, which have approximately the same velocity, but are nearly 2,000 times heavier than the electrons.

Let us assume that the solar plasma carries no magnetic field, and that the particles undergo a specular reflection at the magnetopause. Under these conditions the momentum transferred in each particle reflection is,

$$mV \cos \psi - (-mV \cos \psi) = 2mV \cos \psi \qquad (3.5\text{-}1)$$

where m is the mass, V the velocity, and ψ the angle to the normal of the impinging solar protons (Figure 3.5-I). The number of these particles striking per second a unit area of the magnetopause is,

$$NV \cos \psi \qquad (3.5\text{-}2)$$

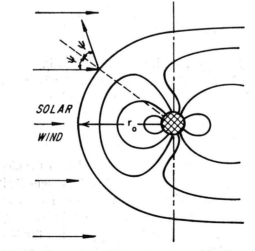

FIGURE 3.5-I   The formation of the magnetopause by the sweeping action of the solar wind

where N is the number of protons per unit volume of the plasma cloud. Thus the pressure (change of momentum per unit time and per unit area) on the magnetopause is,

$$P_H = (2mV \cos \psi)(NV \cos \psi) = 2mNV^2 \cos^2 \psi \qquad (3.5\text{-}3)$$

The balancing pressure, on the other hand, due to the magnetic field H inside the magnetopause, is,

$$p_H = \frac{H^2}{8\pi} \qquad (3.5\text{-}4)$$

where H has approximately twice the intensity that the undisturbed magnetic field would have had at the same distance. The reason for doubling the value of the magnetic field is that due to the compression of the field we have piled up many more field lines near the magnetopause than there would normally be at this distance. Thus if r_0 is the equatorial distance of the magnetospheric boundary, the magnetic field H inside the magneto-pause at the equator is related to the ground magnetic field H_{ge} by the expression,

$$H = 2H_{ge}\left(\frac{R_0}{r_0}\right)^3 \qquad (3.5\text{-}5)$$

Combining now equations (3.5-3), (3.5-4) and (3.5-5), and assuming that at the equator $\psi \simeq 0$, we obtain the relation,

$$\frac{4H_{ge}^2}{8\pi}\left(\frac{R_0}{r_0}\right)^6 = 2mNV^2 \qquad (3.5\text{-}6)$$

which for $N \simeq 2 - 10$ protons/cm^3, $V \simeq 300\text{--}700$ km/sec, and H_{ge} $= 0.3$ Gauss yields,

$$r_0 = \left(\frac{H_{ge}^2}{4\pi mNV^2}\right)^{1/6} R_0 \simeq 10R_0 \qquad (3.5\text{-}7)$$

Because of the 1/6 power, (3.5-7) is not very sensitive to changes in the values of N and V. Also a factor of 2 by which certain authors reduce the pressure of the solar particles because of charge separation at the magneto-pause (protons have a larger cyclotron radius (3.3-3) and therefore penetrate deeper before they turn back) changes very little the above result.

Thus in the direction of the sun the magnetosphere extends only up to about 10 earth radii. This distance can vary from 7 to $12R_0$ as the flux of the solar particles changes with the prevailing level of solar activity. A complete solution of the problem in three dimensions, which would define the entire boundary of the magnetosphere, is an extremely difficult task and has not been accomplished yet in its full generality. Some more restricted forms of the problem, however, have been solved by many investigators using different simplifying approximations.

In their early work Chapman and Ferraro had the impression that the geomagnetic cavity was a transient phenomenon which occurred only after the eruption of a large solar flare. At all other times they thought the earth's magnetic field resumed the old configuration of the undistorted dipole field. Twenty years later, Ludwig F. Biermann (1951) suggested that there is a continuous outflow of high-energy charged particles from the sun which is only intensified during periods of strong solar activity. Prof. Biermann based his theory on the fast velocities with which some inhomogeneities move along the type-I tails of the comets. Type-I tails (Figure 3.5-II) are thin and straight in the antisolar direction and consist mostly of charged particles. Radiation pressure from the sun, which keeps the thick, slightly curved, type-II tails of the comets (Figure 3.5-II) pointing away from the sun, could not produce the very fast movement of the visible inhomogeneities seen in the type-I tails. To explain these very high velocities Biermann proposed the existence of a continuous outward flux of high energy charged particles from the sun. Biermann's original explanation has been modified in recent years, but his original theory was the first suggestion of a continuous corpuscular flux from the sun. This has now been verified experimentally with space probes and the continuous stream of the energetic particles from the sun is called the *solar wind*.

Combining the theories of Chapman and Ferraro, and Biermann, we conclude that the geomagnetic cavity is a permanent phenonmenon. The magnetic cavity responds to changes in the intensity of the solar wind, but the dimensions of the cavity they practically never vary by more than a factor of 1.5 from their average values. During periods of high solar activity the solar wind blows more violently, causing significant changes in the magnetosphere, which also produce noticeable changes of the magnetic field on the ground. Thus satellite measurements beyond the boundary of the magnetosphere together with ground observations of the geomagnetic activity have yielded an empirical expression which relates the velocity of the solar wind V to the index ΣKp of the magnetic activity on the ground. This relation is,

$$V \,(\text{km}) \simeq 330 + 8.44 \Sigma Kp \tag{3.5-8}$$

In Chapter 5 and especially in Section 5.4 we will see that the velocity of the solar wind is always much higher than both the sound velocity and the Alfvén velocity of the interplanetary plasma. Consequently, as the solar wind streams by the magnetic cavity with supersonic speeds, it produces a magnetohydrodynamic shock wave, often called the *bow shock*, on the solar side of the magnetopause. To get a better picture of this effect

FIGURE 3.5-II   The type-I and type-II tails of a comet seen in August 1957.
(Courtesy Mount Wilson and Mount Palomar Observatories)

one can reverse the process and consider the earth with its magnetosphere moving at supersonic velocities through a stationary plasma and creating the well-known, approximately hyperbolic, bow wave.

The region between the edge of the shock wave and the magnetopause is called the *magnetosheath*. This is a transition region where a turbulent interaction of solar wind particles and magnetic fields takes place. It extends from $10R_0$ to $14R_0$ in the subsolar region and from $14R_0$ to $23R_0$ at $90°$ to the sun-earth direction. These of, course, are only average dimensions, and they might vary by as much as 25% depending on the strength of the solar wind. The geomagnetic field has an intensity of $\sim 50\,\gamma$ ($1\,\gamma = 10^{-5}$ Gauss) inside the magnetopause and drops abruptly to $\sim 10\,\gamma$ when we cross the thin (a few hundred kilometers at the subsolar point) boundary of the magnetosphere. In the magnetosheath the magnetic field shows sharp, irregular fluctuations both in intensity (0 to $10\,\gamma$) and in direction. Beyond the shock wave front, we find the relatively steady magnetic field of the interplanetary space. This field has an intensity of about $6\,\gamma$ and actually it is the magnetic field of the sun carried away by the plasma of the solar wind.

The magnetic field of the earth under the sweeping action of the solar wind forms a magnetic tail in the antisolar direction. Thus behind the earth the magnetopause becomes a cylindrical surface. The radius of the magnetic tail R_t is approximately $22R_0$ and remains the same for at least $100R_0$. As seen from Figure 3.5-III, the magnetic field of the tail is actually the magnetic field of the polar caps which has been swept back by the solar wind. The incoming field lines in the northern half and the outgoing field

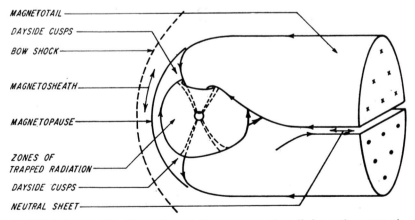

MAGNETOTAIL
DAYSIDE CUSPS
BOW SHOCK
MAGNETOSHEATH
MAGNETOPAUSE
ZONES OF TRAPPED RADIATION
DAYSIDE CUSPS
NEUTRAL SHEET

FIGURE 3.5-III The formation of the geomagnetic tail from the magnetic field lines of the polar caps which are swept in the antisolar direction by the solar wind

lines in the southern half of the magnetic tail are separated by a plane layer where the intensity of the magnetic field drops essentially to zero. This neutral layer has a thickness of about 1,000 km and is called the *neutral sheet*. On several occasions satellites have detected in the neutral sheet weak magnetic fields normal to the neutral plane. This suggests that the parallel but opposite field lines on either side of the plane not only neutralize each other, but occasionally they combine to form loops, like the symbol of infinity, inside the neutral sheet.

Using the concept of the conservation of the total magnetic flux, we can compute the magnetic field H_t in the magnetic tail. We have already obtained in Section 3.2 the total magnetic flux through each one of the polar caps. Using (3.2-23) and the fact that the ground magnetic field at the equator is $H_{ge} = M/R_0^3$, this flux can be written in the form,

$$F = 2H_{ge}\pi R_0^2 \sin^2 \theta_0 \qquad (3.5\text{-}9)$$

This should be equal to the total magnetic flux through the corresponding half of the magnetic tail, i.e.,

$$F = \frac{1}{2}\pi R_t^2 H_t \qquad (3.5\text{-}10)$$

Equating (3.5-9) and (3.5-10) we find,

$$H_t = 4H_{ge}\left(\frac{R_0}{R_t}\right)^2 \sin^2 \theta_0 \qquad (3.5\text{-}11)$$

Satellite measurements in the magnetic tail have found that $R_t \simeq 22R_0$ and $H_t \simeq 16\,\gamma$. Introducing these values, and $H_{eg} = 0.3$ Gauss $= 3 \times 10^4\,\gamma$, in (3.5-10) we obtain,

$$\theta_0 = \sin^{-1}\left\{\frac{R_t}{2R_0}\left(\frac{H_t}{H_{ge}}\right)^{1/2}\right\} \simeq 15° \qquad (3.5\text{-}12)$$

This value is in good agreement with the co-latitude of the horns of the radiation belts and the co-latitude of the auroral zone which mark the end of the polar caps.

The magnetic tail of the earth's magnetosphere has been detected with certainty by satellites orbiting the moon so that it definitely extends beyond half a million kilometers (Ness, 1967). Mariner 4 (Van Allen, 1965) has found no evidence of the tail at $3,300R_0$, whereas observations with Pioneer 7 near $1,000R_0$ were rather ambiguous. Thus it appears that the magnetotail, or a magnetospheric wake behind it after the field lines close, extends for at least several hundred earth radii, i.e., for several million kilometers behind the earth in the antisolar direction. The plane of the neutral sheet is parallel to the plane of the ecliptic but, as seen from Figure 3.5-IV, it is displaced by a few earth radii toward the geomagnetic equator.

On both sides of the neutral sheet satellites have found high fluxes $(10^8–10^9 \, \text{el cm}^{-2} \, \text{s}^{-1})$ of low energy electrons with energies typically of the order of a few keV. This region, which sandwiches the neutral sheet, is called the *plasma sheet* and has a thickness of about $10R_0$. The plasma sheet starts beyond the termination of the belts of trapped radiation, i.e. beyond the last field lines that maintain their nearly dipole shape as the earth rotates, and extends to at least 100 earth radii. This of course means that the charged particles of the plasma sheet are not trapped. Anderson (1965) has also observed "islands" of energetic particles inside the space of the magnetotail. The energies of the electrons in these islands are of the order of 100 keV and the fluxes measured reached $10^7 \, \text{el cm}^{-2} \, \text{s}^{-1}$. It is possible that these islands of energetic electrons are produced by a local pinching of the field lines resulting from the continuous semi-sinusoidal

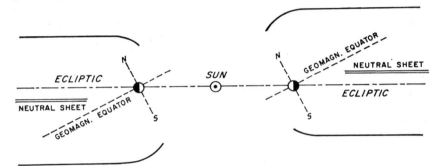

FIGURE 3.5-IV   The movement of the geomagnetic tail relatively to the plane of the ecliptic as the earth orbits around the sun

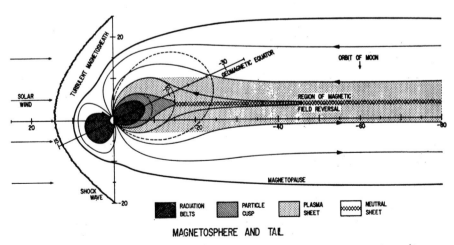

FIGURE 3.5-V   A detailed summary diagram of the earth's magnetic cavity
(Ness, 1969)

motion (flapping) of the geomagnetic tail. A combined diagram of the magnetic sheath, the magnetosphere, and the magnetic tail of the earth, produced by Ness, is shown in Figure 3.5-V.

Measurements with the American Mariner and the Russian Venera space probes have shown that both Mars and Venus lack a magnetic field and, therefore, neither of the two planets is enclosed in a magnetic cavity. Both planets, however, have an ionosphere with a maximum electron density of the order of 10^5 el/cm^3. At first glance it would appear reasonable to expect that the ionosphere of Mars and Venus merges smoothly with the plasma of the interplanetary space, i.e. with the solar wind, without any kind of boundary like the magnetopause of the earth. This is not the case, though, because the solar wind carries with it a weak, but not insignificant, magnetic field which cannot penetrate the ionospheric plasma of Mars and Venus. Thus the two plasma fluids cannot mix and the solar wind is forced to flow around the planets, which in this manner become encapsulated in their own ionospheres. This ionospheric boundary is called the *ionopause* and separates the territorial domains of Venus and Mars from the vastness of the interplanetary space. The ionopause, or *anemopause* as some authors prefer to call it ("anemos" is the Greek word for wind) occurs at an altitude of approximately 1,000 km at the subsolar point.

Jupiter, which we know possesses a strong magnetic field, must be enclosed in a huge magnetic cavity similar in shape to the magnetic cavity of the earth, but much larger in size. Its radial distance at the subsolar point can be obtained from (3.5-7) by changing the particle density of the solar wind as the square of the distance, keeping the velocity of the solar wind nearly constant, and taking the magnetic field on the surface of Jupiter to be approximately 10 times stronger than the magnetic field on the surface of the earth. The result is that the magnetopause of Jupiter at the subsolar point occurs at a distance of 30 to 50 Jovian radii, and since the radius of Jupiter is about one tenth the radius of the sun, it follows that the magnetic cavity of Jupiter extends to several solar radii in the direction of the sun.

3.6 Bibliography

A. Books for Further Studies

1. *Geomagnetism*, S. Chapman and J. Bartels, Clarendon Press, Oxford, England, 1940.
2. *Physics of Geomagnetic Phenomena*, ed. by S. Matsushita and W. H. Campbell, Academic Press, New York, N. Y., 1967.
3. *The Radiation Belts and the Magnetosphere*, W. N. Hess, Blaisdell Publ. Co., Waltham, Mass., 1968.

4. *The Adiabatic Motion of Charged Particles*, T. G. Northrop, Interscience, New York, N. Y., 1963.
5. *Physics of Fully Ionized Gases*, L. Spitzer, Interscience, New York, N. Y., 1963.
6. *Introduction to the Physics of Space*, B. Rossi and S. Olbert, McGraw-Hill, New York, N. Y., 1970.
7. *Magnetospheric Physics*, ed. by D. J. Williams and G. D. Mead, Am., Geophys. Union, Washington, D. C., 1969.
8. *Physics of the Magnetosphere*, ed. by R. Carovillano, J. McClay, and H. Radoski, Reidel Publ. Co., Dordrecht, Holland, 1968.
9. *Earth's Particles and Fields*, ed. by B. M. McCormac, Reinhold Publ. Co., New York, N. Y., 1968.
10. *Radio Astrophysics*, A. G. Pacholczyk, W. H. Freeman and Co., San Francisco, Calif., 1970.
11. *Space Physics*, Sir Harrie Massey, Cambridge University Press, Cambridge, England, 1964.
12. *Space Physics*, R. S. White, Gordon and Breach, New York, N. Y., 1970.
13. *Solar-Terrestrial Physics*, ed. by J. W. King and W. S. Newman, Academic Press, New York, N. Y., 1967.
14. *Introduction to Space Science*, ed. by W. N. Hess and G. D. Mead, Gordon and Breach, New York, N. Y., 1968.
15. *Research in Geophysics*, ed. by H. Odishaw, MIT Press, Cambridge, Mass., 1964.
16. *Satellite Environment Handbook*, ed. by F. S. Johnson, Stanford University, Stanford, Calif., 1965.

B. Articles in Scientific Journals

Anderson, K. A., Energetic electron fluxes in the tail of the geomagnetic field, *J. Geophys. Res.*, **70**, 4741, 1965.

Biermann, L., Kometenschweife und solare Korpuskularstrahlung, *Z. Astrophys.*, **29**, 274, 1951.

Chapman, S. and V. C. A. Ferraro, A new theory of magnetic storms, *Terr. Magn. Atm. Elec.*, **36**, 171, 1931.

Christophilos, N. C., The Argus experiment, *J. Geophys. Res.*, **64**, 869, 1959.

Ginzburg, V. L. and S. I. Syrovatskii, Cosmic magnetobremsstrahlung, *Ann. Rev. Astron. Astrophys.*, **3**, 297, 1965.

Hamlin, D. C. *et al.*, Mirror and azimuthal drift frequencies for magnetically trapped particles, *J. Geophys. Res.*, **66**, 1, 1961.

Krimigis, S. M. *et al.*, Trapped energetic nuclei $z \geq 3$ in the earth's outer radiation zone, *J. Geophys. Res.*, **75**, 4210, 1970.

McIlwain, C. E., Coordinates for mapping the distribution of magnetically trapped particles, *J. Geophys. Res.*, **66**, 3681, 1961.

Ness, N. F., Earth's magnetic field: a new look, *Science*, **151**, 3714, 1966.

Ness, N. F. *et al.*, Observations of the earth's magnetic tail and neutral sheet at 510,000 kilometers by Explorer 33, *J. Geophys. Res.*, **72**, 927, 1967.

Ness, N. F., The geomagnetic tail, *Rev. of Geophys.*, **7**, 97, 1969.

Singer, S. F., A new model of magnetic storms and aurorae, *Trans. AGU*, **38**, 175, 1957.

Van Allen, J. A., Absence of 40-KeV electrons in the earth's magnetospheric tail at 3300 earth radii, *J. Geophys. Res.*, **70**, 4731.

Warwick, J. W., Radiophysics of Jupiter, *Space Science Reviews*, **6**, 841, 1967.

CHAPTER 4

THE ACTIVE SUN

4.1 Introduction

The sun is a star of mass $M = 1.99 \times 10^{33}$ gr, radius $R = 6.96 \times 10^{10}$ cm, and effective temperature $T = 5750°K$. The total energy radiated by the sun per second, i.e. its *luminosity L*, is,

$$L = 4\pi R^2 \sigma T^4 = 3.9 \times 10^{33} \text{ ergs/sec} \qquad (4.1\text{-}1)$$

The sun is a main sequence $G2$ star, approximately 5 billion years old. In many ways it is a very representative star and it is estimated that it will remain essentially in its present state for at least another 5 billion years.

The energy of the sun is produced mostly through the proton-proton nuclear reaction near its center where the temperature is close to $10^7°K$. The carbon nuclear cycle makes also a small contribution to the total energy produced. In both of these processes the end result is that 4 hydrogen atoms fuse together to form a helium atom with the release of approximately 25 MeV $= 4 \times 10^{-5}$ ergs of energy. By comparing this number with the total luminosity of the sun we see that nearly 10^{38} such fusions must take place per second which means that about 6.4×10^{14} gr of hydrogen "burn" per second to helium. In this transmutation 0.7% of the mass becomes energy ($E = mc^2$) and therefore about 4.5 million tons of solar matter are converted every second into energy.

The sun is a gaseous sphere rotating with an average period of 27 days. The word "average" is used because the sun possesses a differential rotation, i.e., its rotational period varies with latitude. The fastest rotation occurs at the equator where the *sidereal* (with respect to the stars) period is very close to 25 days. The rotation slows down with increasing latitude becoming longer than 30 days near the poles. An expression which gives to a good approximation the daily sidereal rotation, in degrees, for different solar latitudes λ is,

$$\phi° \simeq 14.4° - 2.8° \sin^2 \lambda \qquad (4.1\text{-}2)$$

It should be mentioned that for a terrestrial observer the sun appears to rotate with a longer period, called the *synodic period* which, for the equatorial regions, is close to 27 days. The reason of course is that the earth advances by approximately 1 degree per day in its orbit around the sun and after 25 days a point near the equator of the sun needs roughly two more days to reach the new angle of the earth. There is no generally accepted explanation on what causes and maintains the differential rotation of the sun, but it is quite certain that it is responsible for the eleven-year cycle of solar activity. Some scientists have suggested recently that the inner core of the sun might rotate at a much faster rate, but this idea has not gained yet general acceptance.

The sun possesses a rather weak magnetic field which reaches a typical value of a few Gauss on the surface of the sun. Occasionally, however, the solar magnetic field displays transient local enhancements where field intensities can reach values as high as several thousand Gauss. These regions usually become the centers of solar activity which we will discuss in the following sections.

4.2 The Photosphere

The *photosphere* is the layer of the sun from which we receive practically all of the optical emission. One can think of it as a luminous shell at a glowing temperature of nearly 6,000°K. The width of the solar photosphere is determined not by any physical boundaries, but by the degree to which each layer of the solar atmosphere is transparent or opaque to the optical rays. The region above the photosphere is practically transparent to the optical rays which originate in the photosphere. On the contrary the layers below the photosphere are opaque and therefore not accessible to optical observations. The layers that are neither very transparent nor very opaque form the photosphere. Naturally the boundaries of the photosphere are not very sharp because the transition from transparency to opaqueness occurs gradually, and the region where it occurs depends to some extent on the wavelength of the emitted radiation. The thickness, however, of the solar photosphere in white light is only of the order of several hundred kilometers, i.e., less than 0.1 % of the radius of the sun. Hence the photosphere, which emits practically all of the visible solar radiation, is an extremely thin layer of the sun.

The energy emitted by the photosphere is produced in the core of the sun through the conversion of hydrogen to helium and is transferred to the bottom of the photosphere by convection. Inside the photosphere, however, the energy passes from layer to layer in the form of radiation (*radiative transfer*) which at the temperature and density of the photosphere is a much more effective process than convective transfer. As the radiation passes through a thin layer, it loses a small amount which is absorbed by the mass of the layer, but at the same time it also gains a small amount because some radiation is emitted by the matter of the layer. Thus the radiation might end up with a net gain or a net loss (or even both if one considers separately different wavelength regions) after passing through this layer. By considering the photosphere as a series of many such thin layers, we can integrate the above-mentioned gains and losses through the entire thickness of the photosphere to obtain the characteristics of the radiation which is finally emitted from the top of the photosphere. The relation, which describes this entire process is called the *equation of radiative transfer*. This equation, which has an application to many astrophysical problems, is derived in Appendix I, where we also define the different relevant parameters.

In the solar photosphere, most of the energy absorbed in the optical domain is used for freeing electrons which are attached to neutral hydrogen atoms. This is called *negative ion absorption*,

$$H^- + h\nu = H + e^- \tag{4.2-1}$$

and the corresponding absorption coefficient $\varkappa_\nu$ varies by less than a factor of 2 over the entire optical region of the spectrum where most of the solar radiation is emitted. Thus it is quite appropriate to use an average value of the absorption coefficient for the entire spectrum,

$$\varkappa_\nu = \bar{\varkappa} \tag{4.2-2}$$

and treat the photosphere of the sun as *grey matter*, i.e., matter which has an absorption coefficient independent of wavelength. It is also justified to assume that the solar photosphere is in both *local thermodynamic equilibrium* and *radiative equilibrium* (see Appendix I) so that the solution of the equation of radiative transfer is given by the *Eddington approximation* (A-35),

$$I(\theta, 0) = \frac{\sigma T_e^4}{2\pi}\left(1 + \frac{3}{2}\cos\theta\right) \tag{4.2-3}$$

where $I(\theta, 0)$ is the intensity of the solar radiation emitted from the top of the photosphere ($\tau = 0$) at an angle θ to the vertical. T_e is the effective temperature of the sun. Equation (4.2-3) shows that $I(\theta, 0)$ will decrease as θ varies from $0°$ to $90°$. The relative magnitude of $I(\theta, 0)$ with respect to the intensity at the center of the solar disc $I(0, 0)$ is given by the expression,

$$\frac{I(\theta, 0)}{I(0, 0)} = \frac{2}{5} + \frac{3}{5}\cos\theta = 1 - u + u\cos\theta \tag{4.2-4}$$

where $u = 0.60$. Equation (4.2-4) says that the sun is brighter at the center of the disc ($\theta = 0$) than at the limb ($\theta = 90°$) where it loses more than half of its brightness. This phenomenon is called *limb darkening*. The value of u obtained through actual measurements is $u = 0.56$ which means that the theoretical result of the Eddington approximation ($u = 0.60$) is in very good agreement with the observations.

A pictorial explanation of the limb-darkening effect is given in Figure 4.2-I. Optical rays from the sun cannot originate any deeper than a few optical depths (see Appendix I) below the surface of the photosphere. Near the limb, where the rays to the observer propagate along a slant direction to the radial, this depth is reached at a shallower layer than near the center of the solar disc. As seen from (A-30), however, the temperature of the photosphere increases with depth, so that the shallower layers are cooler and therefore less bright, since the intensity of radiation varies as T^4. Hence a decrease of 20% in the temperature, i.e., a change of about 1,000°K, will reduce the brightness by more than a factor of 2.

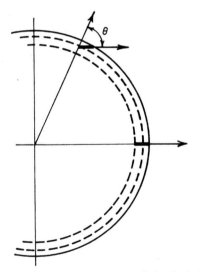

FIGURE 4.2-I   Schematic illustration of the limb-darkening effect

The photosphere, as we have mentioned earlier, is a very thin layer in radiative equilibrium which means that the energy exchange among the photospheric layers is accomplished primarily through radiation. Below the photosphere, however, there is a thick zone in *convective equilibrium,* i.e., a region where the transfer of energy is accomplished mostly through convection. The thickness of this zone is substantial, extending for about 0.2 solar radii below the photosphere. Convective flow usually occurs along vertical cells in which the hot gases rise through the middle and the cooler gases sink back along the rim of the cell. For this reason when seen from above, these cells appear to have bright, hot cores and dark, cool edges. In the case of the sun the upper end of this cellular flow pattern penetrates into the lower photosphere and therefore it can be seen from the earth to divide the visible disc of the sun into nearly a million small polygonal cells. This "mosaic" appearance of the photosphere is called *photospheric granulation* or *solar granulation*. The typical size of the granules ranges between 1,000 and 3,000 km. The narrow dark edges that delineate the bright granules have a fairly uniform width of a few hundred kilometers and form a constantly changing network over the disc of the sun. The average life time of these granules is approximately 7 to 10 minutes and movies made with time-lapse photographs show very nicely the continuously changing pattern of the solar granulation.

The disc of the sun seems to be divided also into roughly a thousand large cells by the horizontal velocity of the photospheric gas which changes both in magnitude and direction from one such region to the next. This phenomenon is called *supergranulation* and it is a relatively recent discovery. Its full extent was not realized until 1960 when R. B. Leighton using the Doppler effect in a new spectroheliographic technique was able to record the velocity component of the photospheric gases along the line of sight. These photographs showed nearly zero velocities in the central region of the solar disc which means that the velocity vector of the photospheric gases is essentially horizontal. The average peak horizontal velocity within each cell is 0.3 to 0.5 km/sec. The average size of the supergranules is in the range of 30,000 to 50,000 km and their life times extend into several hours. The fact that these are much larger cells than the cells of the photospheric granulation suggests that the supergranules are produced by convective instabilities that originate much deeper in the convective zone. They seem to be related also to the overall distribution of the weak magnetic field on the surface of the sun which shows a higher density along the borders of these supergranules (Figure 4.3-I).

4.3 The Chromosphere and the Corona

The region above the photosphere is called the *chromosphere*, which in Greek means the "color-sphere". This name comes from the red color of the chromosphere which is due to the predominance of the $H\alpha$-line (the first Balmer line at 6563 Å) of hydrogen. The chromosphere is approximately 10,000 km thick and becomes visible only a few moments before totality during a total eclipse of the sun. As we have seen, the effective temperature of the sun ($\sim 5,750°$K) occurs at $\tau = 2/3$, i.e., approximately 100 km below the surface of the photosphere. At the top of the photosphere the temperature drops to about 5,000°K and continues to decrease to about 4,500°K in the first few hundred kilometers of the chromosphere.

Beyond this minimum the temperature begins to increase, slowly at first and then more rapidly. It reaches values in the 10,000 to 50,000°K range in the first few thousand kilometers and then the *transition zone* occurs, where the temperature rises steeply from 50,000 to 500,000°K in probably less than a thousand kilometers. This sharp increase in temperature represents one of the most complex and most intriguing problems of solar physics. Above this region the temperature continues to increase but at a much slower rate reaching finally at an altitude of 10,0000 to 20,000 km a temperature of about $1.5 \times 10^{6}°$K which is the nearly constant temperature of the solar corona.

The first 1,000 km of the chromosphere are a fairly uniform layer, but above this height the chromosphere becomes a very inhomogeneous region because it consists of many small jets shooting out into the corona. These projections are called *spicules* and in general have the shape of a rather long cylindrical cone approximately 1,000 km thick and about 5,000 km tall. The spicules continuously rise and fall and their average lifetime is close to 5 minutes. At any instant of time, several hundred thousand spicules are present, but they are not distributed over the solar disc as uniformly as the photospheric granulation. Spicules seem to concentrate more along the lines of the chromospheric network which corresponds to the supergranulation pattern at the photospheric level (Figure 4.3-I). The

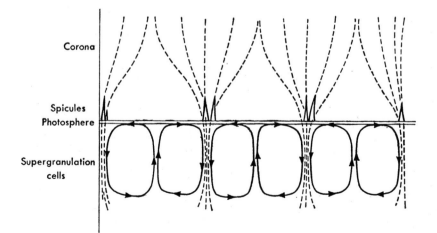

FIGURE 4.3-I The concentration of the spicules and of the solar magnetic field at the borders of the super-granulation (Tandberg-Hanssen, 1966)

spicules are actually jets of relatively cold and dense chromospheric gas ($T \sim 10^{4\circ}$K, $N_e \sim 10^{12}$ el/cm^3) rising with a velocity of ~ 30 km/sec into the much hotter ($T = 10^5$–$10^{6\circ}$K), but also considerably less dense ($N_e \sim 10^{10}$), lower edge of the solar corona.

The weak magnetic field of the sun seems to concentrate also along the chromospheric network and to run in general along the axis of the spicules (Figure 4.3-I) keeping the rising jets together and preventing their quick dispersion into the hot corona. On their top part, spicules might be covered with a thin layer which is at an intermediate stage between the cold, dense core of the spicules and the hot, thin gas of the surrounding medium. This thin coating is probably responsible for the Helium line emission (He I at 584 Å and He II at 304 Å) while the denser and colder material of the spicules is responsible for the emission of the strong H and K lines of Ca II, and of the Balmer lines of hydrogen. The microwave radio emission is also produced by the matter in the spicules while the x-rays and much of the ultraviolet emission is taking place in the coronal gas between and above the spicules.

The solar corona begins essentially in the region between the spicules and extends outwards merging ultimately with the interplanetary medium. The temperature of the corona is approximately $1.5 \times 10^{6\circ}$K and changes very slowly with distance from the sun. The coronal gas at this high temperature is fully ionized and consists essentially of electrons and protons. An empirical expression which is often used for the electron density profile of the quiet corona is the Baumbach-Allen formula,

$$N_e = 10^8(1.55\varrho^{-6} + 2.99\varrho^{-16})\ \text{el/cm}^3 \qquad (4.3\text{-}1)$$

in which $\varrho = r/R_0$ is the ratio of the radial distance r from the center of the sun to the radius R_0 of the photosphere. The electron density of the corona over an active region has approximately the same profile but multiplied by a factor of 10 or 20. The Baumbach-Allen formula applies only to values of ϱ less than about 5, because at larger distances, the density of the solar corona falls off very nearly like the square of the distance, i.e., like ϱ^{-2}.

The sharp rise of the temperature in the transition zone and the heating of the corona to temperatures several hundred times higher than the effective temperature of the sun, represent a very difficult problem which has not been worked out yet in its full complexity. The basic process, however, can be described in the following simple terms. Sound (acoustic) waves are generated by turbulence and convection below the photosphere and they start moving up carrying only about 1 % of the total energy flux. This energy is ultimately dissipated by the sound waves near the transition zone, producing a large increase in the temperature of the medium. The energy density (energy per unit volume) of the sound waves is,

$$E_s = \frac{1}{2} \varrho V_t^2 \tag{4.3-2}$$

where ϱ is the density of the medium, and V_t the *velocity amplitude* of the waves. V_t is equal to the amplitude times the frequency of the waves and represents the average oscillating velocity of the particles in the wave. These compression waves propagate with the speed of sound V_s, which is given by the well known expression,

$$V_s = \left(\frac{\gamma k T}{m}\right)^{1/2} \tag{4.3-3}$$

Hence the energy flux of the sound waves, which is equal to the energy density times the velocity of propagation is,

$$F_s = E_s V_s = \frac{1}{2} \varrho V_t^2 \left(\frac{\gamma k T}{m}\right)^{1/2} \tag{4.3-4}$$

In the photosphere and the lower chromosphere the temperature T and therefore the speed of sound V_s remain fairly constant. The density ϱ, however, decreases rapidly by several orders of magnitude and in order to conserve F_s (assuming zero loss) V_t must keep increasing in proportion to $\varrho^{-1/2}$. Thus ultimately V_t will reach values comparable to V_s, at which point the sound waves will form a shock wave and will rapidly dissipate their energy into the medium. An analogous physical process is the cracking of the whip. Because the whip becomes progressively slimmer (the same as decreasing ϱ), the velocity amplitude of the wave that travels the length of the whip keeps increasing and at the very tip of the whip it often reaches supersonic values producing the familiar cracking sound.

It is reasonable to assume that most of the energy of the shock waves will be dissipated in one scale height H (1.4-4), which is the height interval in which the density of the solar corona changes by nearly a factor of 2. This means that the energy flux F_s will be equal to the energy W_D dissipated per unit time and per unit volume, times the length H over which the flux F_s is dissipated. Hence we have,

$$W_D = \frac{F_s}{H} = \frac{\frac{1}{2}\varrho V_s^3}{H} = \frac{\frac{1}{2}\varrho\left(\frac{\gamma k T}{m}\right)^{3/2}}{\left(\frac{k T}{mg}\right)} \simeq 10^8\, \varrho T^{1/2}\ \text{erg cm}^{-3}\ \text{sec}^{-1} \tag{4.3-5}$$

7*

Note that in the chromosphere, where the acoustic waves change to shock waves ($V_t \sim V_s$), the acceleration of gravity is $g = 2.7 \times 10^4$ cm/sec^2. Equation (4.3-5) shows that the dissipation of the energy carried by the sound waves provides a very substantial energy source in the chromosphere. Part of this energy, however, will be lost through radiation and will not be available for building up the temperature of the medium.

The major portion of the energy radiated away by the chromospheric gas is emitted in the Lyman continuum (bound-free transitions) where the energy emitted per unit volume and per unit time is given by the semi-empirical relation,

$$W_R = 5.5 \times 10^{-22} N_e^2 T^{-1/2} \text{ erg cm}^{-3} \text{ sec}^{-1} \qquad (4.3\text{-}6)$$

By equating (4.3-5) and (4.3-6) and assuming complete ionization so that $\varrho = mN_p = mN_e$, we find that the radiation losses can balance the dissipated energy up to a height where $N_e \simeq 3 \times 10^5 T \simeq 10^{10}$ el/cm^3. At greater altitudes, however, as N_e decreases rapidly the radiation losses cannot keep up with the dissipated energy and the excess energy appears as a sharp rise in the temperature of the gas. Finally the temperature reaches a level close to 10^{6}°K at which point the thermal conductivity of the plasma becomes so high that all the dissipated heat can readily be dispersed by conduction. Due to its very high conductivity (5.2-14) the entire corona is maintained at a nearly uniform temperature.

In a more sophisticated treatment of the problem one must take into consideration also the magnetic field of the sun and replace the simple acoustic waves with the different modes of magneto-acoustic waves with their very complex coupling mechanisms. The result is that the full solution of the problem is too complicated, but the simple process described above gives all the essential elements of the mechanism by which the solar chromosphere and the solar corona are heated up to their respective high temperatures.

Without any special instruments, the corona can be observed only during total eclipses. In 1930, however, Bernard Lyot developed the *coronograph* which made it possible to observe the corona at any time. The coronograph is essentially a telescope equipped with a special dark disc which occults the photosphere and allows the viewing of the dim light of the corona. The surface brightness of the corona, however, is only a few millionths of the surface brightness of the solar disc and for this reason these observations are always quite difficult to make.

In the polar regions the corona shows a filamentary structure and some people have suggested that these filaments correspond to the polar magnetic field of the sun. Larger extensions of the solar corona, with dimensions of the order of one or more solar radii, are called *coronal streamers*. Near the minimum of the solar cycle, the corona is cooler ($\sim 1.2 \times 10^{6}$°K) and appears brighter near the equatorial plane, where long streamers can be seen. During the solar maxima, on the other hand, the corona is hotter ($\sim 1.8 \times 10^{6}$°K), nearly twice as dense and almost equally bright around the solar disc. Hotter regions over active sunspots give often a somewhat irregular appearance to the corona during periods of high solar activity.

The optical emission of the corona consists of three components. The first one is the continuum emission due to the Thomson scattering of the photospheric light by the electrons of the corona. These electrons, at $T \simeq 1.5 \times 10^{6}$°K have very high velocities,

$$V = \left(\frac{3kT}{m_e} \right)^{1/2} \simeq 10^9 \text{ cm/sec} \qquad (4.3\text{-}7)$$

and wash out all the Fraunhofer lines of the spectrum through strong Doppler broadening which is given by the expression,

$$\frac{\Delta\lambda}{\lambda} = \frac{V}{c} \qquad (4.3\text{-}8)$$

For $V \sim 10^9$ cm/sec and $\lambda \sim 5,000$ Å, the Doppler broadening of the lines exceeds 100 Å and the lines disappear. This component of the coronal emission is called the *K-corona* and is the dominant component up to a distance of approximately 2 solar radii from the center of the sun.

The second component of the coronal light is due to the diffraction of the sunlight by solid particles (dust) in the space between the sun and the earth. This component is called the *F-corona* or the *inner zodiacal light* because it is the continuation of the faint band of dust-scattered light which occasionally can be seen near the horizon. This diffuse glow is called the zodiacal light because the band usually extends along the plane of the ecliptic which is characterized by the 12 constellations of the zodiac. The *F*-corona becomes the dominant component in the outer corona because the intensity of the *K*-corona decreases rapidly with distance from the sun.

The third component of the coronal light is the sum of all the discrete emission lines of the corona and is called the *E-corona*, or sometimes the *L-corona*. The origin of these lines remained a mystery for a long time and originally their unfamiliar spectra were ascribed to hypothetic new elements, such as the famous *coronium*. This line of thought persisted until the early 40's when Edlen and Grotrian showed that these lines are actually *forbidden lines* from highly ionized metals such as Fe XIV (an iron atom with 13 electrons missing).

The energy required to produce these highly ionized states is of the order of several hundred electron volts as compared to only 13.6 eV required for the ionization of hydrogen. Such energies, however, can easily be found in the corona where, with $T = 1.5 \times 10^{6}°K$, the average kinetic energy of the coronal particles is,

$$E_k = \frac{1}{2} mV^2 = \frac{3}{2} kT = 3.1 \times 10^{-10} \text{ erg} \simeq 200 \text{ eV} \qquad (4.3\text{-}9)$$

and the average energy of a photon (through Wein's displacement law, $T\lambda_{max} \simeq 0.3$) is,

$$E_p = h\nu = \frac{hc}{\lambda_{max}} \simeq \frac{hcT}{0.3} \simeq 10^{-9} \text{ erg} \simeq 600 \text{ eV} \qquad (4.3\text{-}10)$$

In addition, both the electrons and the photons have an energy distribution and therefore one can find photons and electrons with energies even higher than the average values obtained above.

The remaining electrons of these highly ionized atoms can be elevated to a higher energy state through collisions with low energy (a few eV) electrons and photons. The return to the ground state in some of these transitions takes several seconds, while an ordinary atomic transition takes only about 10^{-8} sec. Under laboratory conditions these ions experience many collisions per second and therefore they do not have the time to emit the photons of this slow transition because they are de-excited much sooner by collisions. As a result these lines cannot be seen in the laboratory and for this reason they are called *forbidden lines*. The corresponding excitation levels are called *metastable states* to indicate that electrons remain much longer in these energy levels.

The temperatures of the corona which were deduced from the intensities of the forbidden lines and from the Doppler width of the forbidden lines were found initially to differ by almost a factor of 2. This discrepancy was removed when it was finally realized that *dielectronic recombination* plays an important role in the ionization balance of the solar corona. In this process an electron and an ion recombine by first forming an unstable state with two electrons in excited states.

The solar corona emits also a soft x-ray flux which sometimes is referred to as the *X-corona*. The x-ray spectrum of the quiet corona, which peaks around 20–30 Å, is essentially the integrated flux of many x-ray emission lines. These lines correspond to the very large (several hundred eV) ionization potentials of some of the highly ionized atoms we mentioned in connection with the forbidden lines of the corona. In addition to the emission lines, the hot coronal plasma emits also an x-ray continuum through thermal bremsstrahlung (free-free emission), but because the solar corona is nearly transparent (optically thin) in the x-ray region this contribution is not very important. Free-free emission occurs when free energetic electrons are forced to slow down in their close encounters with ions. The energy lost by such an electron, which is emitted in the form of a photon, depends to some extent on the initial energy of the electron and more critically on the closeness of the encounter. As a result the bremsstrahlung spectrum of the solar corona extends from the x-ray region to the radiowave domain. Naturally free-free emission is also present in the optical region but there it is completely superceded by the *K*-corona which we have already discussed.

The bremsstrahlung emission of the solar corona in the radio domain, which sometimes is referred to as the *R-corona*, is of great interest because it provides information on the temperature and the electron density profile of the solar corona. The corona becomes opaque (optically thick) for radio frequencies below the local plasma frequency of the corona. As a result the size of the radio disc increases as the frequency decreases. Another interesting effect in the radio domain is the *limb brightening* of the radio disc. As seen from Figure 4.3-II, due to refraction effects the radiation from the central region of the disc originates at a deeper layer of the solar corona than the radiation coming from the vicinity of the limb. The temperature, however, is lower in the deeper layers near the chromosphere and therefore the radio brightness $I_v\,dv$, which is proportional to the temperature (see A-43),

$$I_v\,dv = \frac{2kT}{\lambda^2}\,dv \tag{4.3-11}$$

is higher near the limb than near the center of the disc. As we mentioned earlier, in the optical domain we had the opposite effect, i.e., limb darkening, because in the photosphere the temperature increases in the deeper layers.

Beyond the maximum near the limb, the radio brightness of the sun decreases because the corona becomes progressively more transparent. If T_0 is the uniform temperature of a medium and τ its opacity (the integral of the absorption over path length of a ray), the brightness temperature T_b we observe is given to a first approximation by the expression (see A-45),

$$T_b = T_0(1 - e^{-\tau}) \tag{4.3-12}$$

For a very opaque medium, τ tends to infinity and therefore T_b approaches T_0. This is the case of an *optically thick* medium where we only see the top layers and detect their actual temperature. In a rather transparent (*optically thin*) medium, however, τ is a number smaller than 1 and $e^{-\tau}$ has a value closer to 1 than to 0. Hence T_b is considerably smaller than T_0. This is the case in the outer corona where τ is proportional to N_e^2 (2.4-23), and N_e decreases with distance as r^{-6} (4.3-1). Finally when the medium becomes completely transparent both τ and T_b tend to zero, which means that we cannot observe the temperature of a completely transparent medium like we cannot see a totally transparent glass.

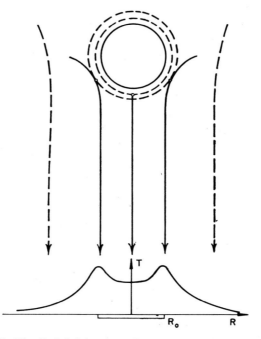

FIGURE 4.3-II   The limb-brightening effect of the solar corona at radio frequencies

In concluding this section it should be pointed out that though the optical emission of the sun never varies by more than 1 %, the radio and x-ray fluxes vary by at least an order of magnitude during the 11 year cycle of solar activity and occasionally can reach peaks which exceed the radio and x-ray background of the quiet sun by several orders of magnitude. These transient enhancements, which we will study in Section 4.6, are called radio bursts and x-ray bursts and usually occur together with the optical flares in an active region on the sun. These centers of activity are usually characterized by a concentration of large sunspots which we will discuss in the next section.

4.4 Sunspots and the Solar Cycle

Sunspots are small, dark, transient spots on the surface of the sun. They can easily be seen with the naked eye by projecting the image of the sun on a white surface. It is believed that Theophrastus of Athens, a pupil of Aristotle, was the first one to observe the sunspots around 300 B.C. The Chinese compiled many naked-eye records of sunspots from the 1st to the 17th century. Sunspots were observed for the first time through a telescope in the year 1611 by several people, including Galileo, in three different countries. Sunspots have a central dark region which is called the *umbra*, and a less dark region which surrounds the umbra and is called the *penumbra*. The umbra is nearly featureless but the penumbra consists of many radial filaments which are believed to be due to roll convection of photospheric matter along the radial magnetic field lines of the sunspot.

A significant contribution in the study of the sunspots and in general of the solar atmosphere, was the discovery of the *Wilson effect* by Alexander Wilson in 1796. Wilson observed that while the sunspots have an essentially symmetric penumbra when they are near the center of the solar disc, they develop an assymmetric penumbra as they move closer to the limb with the narrower side of the penumbra further away from the limb. This is shown diagrammatically in Figure 4.4-I. Wilson himself offered a geometric explanation for this effect by suggesting that sunspots are saucer-shaped depressions on the surface of the sun. Today we know that the Wilson effect is due to a change in the optical depth rather than the physical depth in the region of the sunspot.

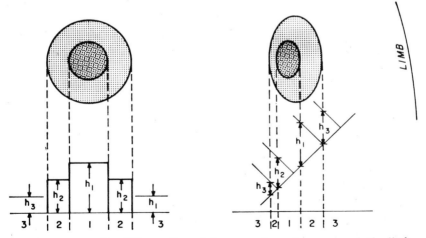

FIGURE 4.4-I The Wilson effect of the sunspots as they approach the limb of the solar disc

Sunspots appear as dark spots on the bright disc of the sun simply because they are cooler and therefore less bright comparing to the rest of the photosphere. But the absorption coefficient $\varkappa$ of the photospheric layers decreases with decreasing temperature and therefore the umbra, which is the darkest and therefore the less hot region, has the lowest absorption coefficient. The absorption coefficient of the penumbra has an intermediate value between the absorption coefficient of the umbra and that of the rest of the photosphere. For simplicity let us assume that we can see up to an optical depth $\tau = 1$, which corresponds to a physical depth $h = \tau/\varkappa = 1/\varkappa$. Since $\varkappa_1 < \varkappa_2 < \varkappa_3$ for the umbra (1), the penumbra (2) and the photosphere (3), it follows that $h_1 > h_2 > h_3$, and therefore the depths to which we can see in the different regions have the physical shape of a soup plate with a rather wide saucer rim. Thus the final effect due to the change of the optical depth is essentially the same like the physical depression which was originally proposed by Wilson.

It was not until 1843 that Heinrich Schwabe announced in Germany that his long sunspot observations had shown that the average number of sunspots on the sun varies with a period of approximately 10 years. Further studies in past records by Rudolf Wolf confirmed the existence of an 11 year sunspot cycle and in 1851 Wolf introduced his *relative sunspot number R* which is given in the relation,

$$R = k(10g + f) \qquad (4.4\text{-}1)$$

where g is the number of sunspot groups on the disc of the sun, f the number of the individual sunspots and k a coefficient assigned to each observing station to assure uniformity in the R numbers obtained by the different stations. In 1855 Wolf became the director of the new Zurich observatory and established there a long tradition of solar observations. For this reason R is also called the *Wolf sunspot number* or the *Zurich sunspot number*. Figure 4.4-II shows the variation of the yearly average of the sunspot number since 1755. Data for the first hundred years were reconstructed from old records by Wolf. In 1969 we passed over the maximum of the 20th cycle of this diagram. The average sunspot cycle lasts 11 years but there are many variations from cycle to cycle and usually the ascending part of the cycle is shorter ($\sim$4.5 years). Around the sunspot maximum, the 12 month average of R approaches, and often exceeds the value of 100, while near the sunspot minimum R has values close to, or even below 10. In more recent years it has been found that all other manifestations of solar activity follow essentially the same 11 year cycle. As a result several other indices of solar activity have been devised, such as the Ottawa index of solar radio emission at 2800 MHz, areas of calcium plages, etc. (Xanthakis, 1969).

Nevertheless the sunspot number still remains the most frequently employed measure of solar activity. It should be mentioned that the cycles of the various indices show considerable differences over the 11 year cycle and their maxima might differ by a year or more. Gnevyshev (1967) has presented data which support his theory of a solar cycle with two maxima, corresponding to high and low latitude activity, which differ by 1–3 years.

Soon after the discovery of the sunspot cycle by Schwabe, Carrington observed that the average latitude of the spots decreases steadily through the solar cycle. When the cycle starts on the ascending mode, there are only a few sunspots around 30° latitude (both north and south). At the maximum, the sunspots are concentrated near a latitude of 15°, and at the end of the cycle there are only a few sunspots left around 5°, while some new ones are reappearing at 30°. Very seldom can sunspots be found at latitudes higher than 45° or closer than a few degrees to the equator. A diagram showing the latitudes of the sunspots during a few solar cycles is shown in Figure 4.4-III. Because of its appearance, it is often called the *butterfly diagram*.

Often the spots appear in pairs with magnetic fields of opposite polarity, i.e., the magnetic field seems to come out of the one spot and enter into the other. The interesting observation is that if the leading spots in the northern

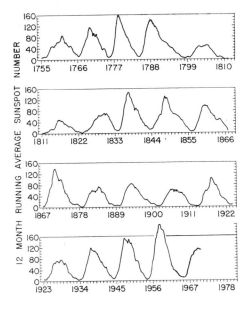

ZÜRICH SUNSPOT NUMBERS

FIGURE 4.4-II The periodic changes of the sunspot number during the last 20 cycles of solar activity

hemisphere have positive magnetic polarity, the leading spots in the southern hemisphere have negative polarity. This arrangement lasts for an entire cycle, but during the following cycle the polarities of the leading spots in the two hemispheres are reversed. For this reason it is probably more accurate to speak of a 22-year solar cycle which includes two 11-year sub-cycles.

An attractive theoretical explanation for the development of the solar cycle, including the latitude effects and the polarity mentioned above, was first proposed by Babcock (1961). Babcock's model (see also Leighton, 1969) assumes that the sun possesses a dipole magnetic field which is confined mostly in a thin ($0.05R_0$) layer below the photosphere. As the equatorial zone of the sun rotates faster than zones of higher latitude, the magnetic field lines are slowly wound around the sun (Figure 4.4-IV) in a manner that increases their flux density at the lower latitudes. Babcock estimated that in three years a point at the equator has gained approximately two full turns over a point at 30° latitude, and that the magnetic field H of the sun at the latitude λ is given by the expression,

$$H = 35.2H_0(n + 3) \sin \lambda \qquad (4.4\text{-}2)$$

where H_0 is the initial magnetic field of the sun (~ 5 Gauss) and $n = 0,1,2,...$ starting from the third year when $n = 0$. It is believed that a magnetic field of about 260 Gauss is sufficient to initiate sunspot activity. This critical

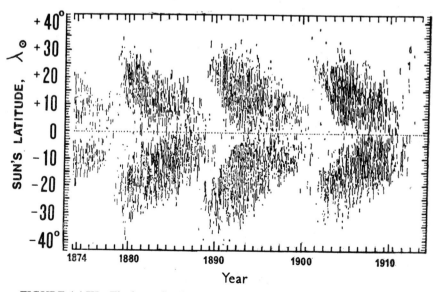

FIGURE 4.4-III   The butterfly diagram which shows the changing of the latitude of the sunspots during the solar cycle (White, 1970)

value of H (taking $H_0 = 5$ Gauss) will occur at a critical latitude λ_c where as seen from (4.4-2),

$$\sin \lambda_c = \pm \frac{1.5}{(n + 3)} \qquad (4.4\text{-}3)$$

Equation (4.4-3) is in agreement with, and explains the butterfly pattern. When the magnetic field lines reach the critical density of ~ 260 Gauss, they start to twist through convection and local irregularities and form ropes or strands of twisted field lines. The flux density of the magnetic field in these ropes can reach values of several thousand Gauss which corresponds to a very high magnetic pressure P_H because,

$$P_H = \frac{H^2}{8\pi} \qquad (4.4\text{-}4)$$

When the pressure of the magnetic field reaches the pressure of the surrounding gas, the ropes achieve magnetic buoyancy and are lifted to the surface where they emerge forming an arch. The area where the magnetic loop emerges and the area where the loop reenters into the photosphere have opposite polarities. This explains nicely the typical pair of sunspots with opposite polarities we mentioned above. As seen from Figure 4.4-IV, the model predicts also correctly that the polarity of the leading sunspot will be reversed when we go from the northern to the southern hemisphere.

The loops of the field lines from a bipolar magnetic region expand slowly into the corona where they meet the field lines of the sun's dipole magnetic

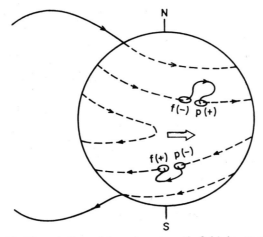

FIGURE 4.4-IV The winding of the solar magnetic field due to the differential rotation of the sun. This process explains why the preceding (p) or leading spots in the two hemispheres have reversed polarities. The same of course is also true for the following (f) spots

field. Eventually a reconnection of field lines takes place which results in the neutralization of the original loops, while large tenuous field loops are liberated and drift into the interplanetary region embedded into the outflowing plasma of the solar wind. Meanwhile, the remaining magnetic field lines under the photosphere pile up from both hemispheres at the equator where eventually they cancel each other. In this way the observable magnetic field of the sun decreases to zero and then it reappears with opposite polarity, i.e. with the north and south poles of the dipole field reversed, to start a new 11-year cycle of solar activity. As we have seen Babcock's theory predicts all the essential features of the 11-year, or better of the 22-year solar cycle, but we still do not understand the mechanism which provides the necessary energy for the differential rotation of the sun, which as we have seen, is the basic cause of the solar cycle.

Sunspots appear at first as small round dark dots 1,500 to 3,000 km in diameter. These infant sunspots are called *pores*. Usually several pores appear simultaneously as a group, but most of these groups disappear in one or two days. Those that survive, grow rapidly in area and separate into a leading and a following group. Both groups are usually dominated by a single large spot which are called the *leading* and the *following* sunspot respectively. The spots develop a penumbra on the 3rd or 4th day and reach a maximum development between the 5th and the 12th day. The leading spot lasts longer, taking usually several weeks to decay and sometimes several months. The following spot, on the other hand, decays faster usually by breaking into several smaller spots. The diameter of the umbra of an average sunspot is 20,000 km while the entire spot, including the penumbra, is 30,000–50,000 km in diameter. Occasionally, however, sunspots have reached much larger dimensions. In a case observed in May, 1964, e.g., the leading spot was 100,000 km long and 56,000 km wide, while the following spot was 150,000 km long and 100,000 km wide. The entire group was over 300,000 km long, i.e., more than 0.2 of the diameter of the solar disc. This gigantic group lasted 99 days, surviving 4 rotations of the sun.

The flux of radiation, integrated over all wavelengths, that is emitted from the center of the umbra F_{um} is about 0.4 of the photospheric flux F_{ph}. Since the total energy emitted is proportional to the 4th power of the effective temperature, this implies that,

$$\frac{T_{um}}{T_{ph}} = \left(\frac{F_{um}}{F_{ph}}\right)^{1/4} = (0.4)^{1/4} = 0.8 \qquad (4.4\text{-}5)$$

from which it follows that,

$$T_{um} = 0.8 \times T_{ph} = 0.8 \times 5,750 = 4,600°K \qquad (4.4\text{-}6)$$

The effective temperature of the penumbra is approximately 5,500°K, which is considerably closer to the effective temperature of the photosphere. Sunspots appear in areas where, as we have seen, there is a strong concentration of solar magnetic field. These strong magnetic fields, which extend to considerable depths below the surface, inhibit the convective transport of energy near the top layers of the convective zone and thus reduce the amount of energy supplied to the photosphere. As a result, a sunspot appears darker because the photosphere in this area is less hot and therefore less bright.

The intensity and polarity of the magnetic fields on the sun are measured using the Zeeman splitting of certain spectral lines. If the field is perpendicular to the line of sight the line splits into three linearly polarized components, while if the field is parallel to the line of sight it splits into two circularly polarized components. By examining the left-handed or right-handed polarization of these circularly polarized components we can tell whether we are dealing with an area of positive or negative polarity.

The magnetic field of a sunspot has its peak strength at the center of the spot and decreases radially approximately as the square of the distance,

$$H(r) = H_m\left(1 - \frac{r^2}{b^2}\right) \tag{4.4-7}$$

where H_m is the peak magnetic field at the center of the sunspot, r the radial distance, and b the value of r at the outer edge of the penumbra. The sunspot field is usually vertical near the center of the umbra and progressively becomes more horizontal as we move further away into the penumbra. The total magnetic flux F_H from a sunspot is typically,

$$F_H = \frac{1}{4} H_m\pi b^2 \tag{4.4-8}$$

and finally if A is the area of the sunspot in millionths of the visible surface of the sun, then H_m can be approximated by the empirical formula,

$$H_m = 3,700\frac{A}{A + 66} \text{ Gauss} \tag{4.4-9}$$

Sunspot groups are divided into three categories: *unipolar* (class-α), *bipolar* (class-β), and *complex* (class-γ). About 90% of all the groups observed belong to class-β, which contains either two spots or two groups of spots of opposite polarity. The remaining 10% belong mostly to class-α which contains either single spots or groups of spots of the same polarity. Only about 1% of all the groups belong to class-γ, which contains groups with many spots of mixed polarity. These complex groups have the highest rate of flare production. There is also a nine-step (class A, B, C, D, E, F, G, H, J) Zurich classification system of sunspot groups which essentially describes the different evolutionary stages that a sunspot group may pass through during its development and decay.

All these detailed classification schemes and the careful counting of the sunspots reflect our efforts to get a handle on the nature of the solar activity. This is a very important task not only from a pure scientific standpoint, but also from a practical point of view because the active sun has an important bearing on our everyday lives on earth and on man's occasional travels into outer space.

4.5   Faculae, Flares and Prominences

In this section we will discuss the different optical manifestations of the active sun. The first sign of solar activity is the appearance of a bright area which, when near the limb of the sun, is brighter than the photosphere even in white light. These bright areas are called *faculae*, or more precisely *photospheric faculae*. Faculae usually engulf a sunspot group, but they become noticeable before the appearance of the sunspots and often survive them by a month or two. The top layers ($\tau \sim 0.2$) of a facula are a few hundred degrees hotter than the photosphere, while the deeper layers ($\tau \sim 1$) are probably somewhat cooler than the corresponding layers of the photosphere. For this reason, photospheric faculae can be seen only near the limb of the sun where, as we have seen in the case of limb darkening, emission originates from the uppermost layers of the photosphere.

These bright areas, however, can be seen in any part of the solar disc in monochromatic pictures of the sun (*spectroheliograms*) taken in the Hα-line of hydrogen or the K-line of calcium. These strong emission lines, however, are produced in the chromosphere and for this reason bright areas observed in this manner are called *chromospheric faculae* or *plages*. This last term is used more frequently and stems from the French word "plage" which means "beach", presumably in this case a beach of bright white sand. The tenuous gas of the chromosphere is essentially transparent for white light and the line emission is too faint to be seen against the bright photosphere. For this reason these hotter regions of the chromosphere can be seen only in the isolated light of a few strong emission lines.

From the temperature structure of the photospheric faculae it follows that these regions are not in radiative equilibrium. Most probably they represent an intensification of the hydromagnetic process, which heats the chromosphere, due to a local increase of the magnetic flux. Plages and faculae are manifestations of the same phenomenon at different heights of the solar atmosphere. Plages usually outlive the corresponding faculae by a few days or a few weeks.

A flare is a sudden local increase in the brightness of the solar surface which lasts for nearly one hour. Flares appear in an active region, i.e., a plage area with sunspots. The eruption of a solar flare is the culmination of the activity that has been mounting up in the sunspot region. A flare is the optical effect produced by the sudden release of tremendous amounts of energy in the upper chromosphere or the lower corona. This explosion, which probably takes place above the layers where the optical flare appears, produces also a strong outburst of x-ray and radio emission which we will discuss in the next section.

When seen in projection at the limb of the sun, flares appear to have different shapes (mounds, spikes, cones, loops, etc.) that start in the chromosphere and extend into the lower corona. Typical heights range between 2,000 and 20,000 km. Large flares can be seen in the white light, but flares are usually observed in the Hα-line where they can be seen much better. Flares are classified in 5 categories of *importance* according to their area at the time of their maximum brightness. The areas are naturally corrected for projection effects due to the position of the flare on the solar disc.

The five classes of flares are designated as: Imp. $1-$, 1, 2, 3, and $3+$. Flares of importance $3+$ are the largest and flares of importance $1-$, which are also called *subflares*, are the smallest. The area of a flare is usually measured in millionths of the visible hemisphere ($2\pi R^2$) of the sun. One millionth is equal to 3×10^{16} cm^2, which is very nearly the area of a circular region 1,000 km in radius. As seen from Table 4.5-I, which gives the area range for the five classes of flares, an Imp. $3+$ flare will brighten up an area more than 70,000 km in diameter. Flares are also classified according to their brightness in three groups: *faint* (F), *normal* (N), and *bright* (B), so that flares are designated as 1N, 1B, etc.

TABLE 4.5-I

Importance	Corrected Area in millionths
1−	<100
1	100–250
2	250–600
3	600–1200
3+	>1200

There are several observatories around the world which patrol continuously the sun and provide flare data. These lists do not always agree in the classification of a given flare because the flare area which must be measured has often a very complex shape that changes rapidly with time. Flares always occur in active regions especially near complex (type-γ) sunspot groups. Flares often appear in the region between two parallel rows of sunspots of opposite polarity. They almost never occur over the umbra of a sunspot, though occasionally they might appear over the penumbra of a spot. The average distance between flares and sunspots is of the order of 10,000 to 30,000 km. Most flares occur in the 10° to 20° latitude range where we also have the highest concentration of sunspots during solar maximum. Some active regions might never produce a flare while others, especially some long lasting ones, might produce more than 10 flares. As many as 100 flares per active region have been reported for a few very exceptional cases. Approximately 80% of all the flares recorded are of importance 1− (subflares). Among the other classes the distribution is: Imp. 1 ($\sim$77%), Imp. 2 ($\sim$20%), Imp. 3 ($\sim$2%), and Imp. 3+ ($\sim$1%). The quoted figures vary somewhat during the solar cycle.

The brightness of a flare rises rapidly to its peak intensity ($\sim$5 min. for flares of Imp. 1), stays for a short period at the maximum and then slowly decays back to normal. The average duration of a flare is: Imp. 1− ($\sim$10 min.), Imp. 1 ($\sim$20 min.), Imp. 2 ($\sim$40 min.), Imp. 3 ($\sim$60 min.), and Imp. 3+ ($\sim$150 min.). Near solar maximum there are 4–6 flares of importance $\geq$1 per day, while at solar minimum we can observe approximately only one such flare every other day.

The area of a typical flare event is $\sim$250 millionths of the visible surface of the sun, i.e. $\sim$7.5 $\times$ 10^{18} cm². The thickness of the region where the flare event takes place is estimated to be of the order of 4,000 km, and therefore the volume of the events is about 3 $\times$ 10^{27} cm³. The total energy released during a typical flare event is approximately 10^{31} ergs, and therefore the energy density available in the flare region must be of the order of 3 $\times$ 10^3 ergs/cm³. The thermal energy density in the chromosphere is,

$$E_T \sim NkT \sim 10^{13} \times 10^{-16} \times 10^4 \sim 10 \text{ ergs/cm}^3 \qquad (4.5\text{-}1)$$

which obviously is totally inadequate. The same is also true for the corona and for the kinetic energy in the vertical motion of the spicules. As a result the magnetic field seems to be the only energy source available that can fulfill this very high energy requirement. With $H \sim$ 300 Gauss, the energy density of the magnetic field is,

$$E_H = \frac{H^2}{8\pi} = 3.6 \times 10^3 \text{ ergs/cm}^3 \qquad (4.5\text{-}2)$$

which is entirely adequate and at the same time very realistic, since magnetic fields of several hundred Gauss can easily be found in a sunspot region.

A high concentration of energy, however, is not the only requirement. Any workable mechanism must be able to release also this energy in a time interval of only a few minutes. One of the earliest attempts to use magnetic field energy for a flare event was proposed by Gold and Hoyle (1960). Their model consisted of two nearly parallel arches of oppositely twisted magnetic field lines. Axial currents pull the two arches together, and as the one bundle of field lines penetrates into the other, the opposite magnetic fields annihilate each other releasing sufficient energy to account for even the largest flares. Such arches of twisted magnetic field lines are not unrealistic bacause they are regularly produced, as we have seen, by the differential rotation of the sun and the strong convection irregularities under the photosphere.

The currently most favored theory was originally proposed by Sweet (1958) and Severny (1958) but was expanded and modified subsequently by many other authors. The magnetic field of a rather complex arrangement of magnetic poles is initially pulled outwards by the solar wind. As it expands it can form one or more neutral points, or a neutral line. A neutral point, i.e., a point of zero magnetic field, is produced when field lines of opposite polarity converge into a point and neutralize each other. Figure 4.5-I shows a neutral point at the center of the diagram. One can think also of other models (see Section 7.2), where instead of a neutral point we have a neutral line or even a neutral plane.

When the energy density of the magnetic field becomes much higher than the thermal energy $\left(\dfrac{H^2}{8\pi} \gg NkT \right)$, a plasma instability develops which produces a strong pinching effect of the magnetic field (wide, horizontal arrows in Figure 4.5-I). The pinch effect causes a cataclysmic implosion of the solar plasma at the neutral point. Densities of the order of 10^{14} cm^{-3} and temperatures of the order of 10^7–10^{8}°K can be reached in a matter of seconds in the small volume at the center of the implosion. This compression is followed by an explosion, the shock front of which can be seen travelling over most of the visible surface of the sun in time lapse photography made in the monochromatic light of the Hα-line. The field lines crossing at the neutral point, break and reconnect (dotted lines in Figure 4.5-I) in a way which resembles the extended rubber bands of a sling shot. These lines quickly snap back (thin, vertical arrows) accelerating particles both upwards toward the corona, and downward toward the base of the chromosphere. In this way the collapsing magnetic field is converted into kinetic

8*

and radiative energy and can account in principle for the complex emission processes occurring during a flare event.

Solar *prominences* have been recorded in some very impressive sequences with time-lapse photography. When seen at the limb of the sun, prominences appear as luminous arch-like structures with continuous internal motion. These arches are about 200,000 km long but only a few thousand kilometers thick. When they are projected on the luminous disc of the sun they simply appear as long dark *filaments*. These wiggly filaments, or the equivalent bright arches at the limb, can reach dimensions up to $^1/_4$ of the solar circumference and a few exceptional prominences have reached distances of a million kilometers beyond the limb of the sun. Prominences often show a fine structure in the form of threads, 1,000 km or less in diameter. Prominences are divided into two large groups. The *active prominences* which appear over a sunspot group and for this reason they are also called *sunspot prominences*, and the *quiescent prominences* which are associated with peculiar regions without sunspots or with sunspot groups in their decaying stage.

Quiescent prominences change much slower (hours or days) and last much longer (weeks or months) than the active prominences which might last only a few hours. The temperature of the quiescent prominences is

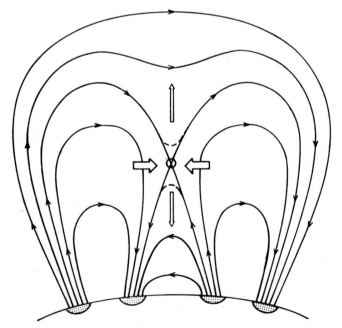

FIGURE 4.5-I   The development of an instability at the neutral point of a moderately complex magnetic field configuration over a sunspot group

about 10,000°K, but their density is more than 100 times the density of the surrounding corona. As a result the gas pressure ($P = NkT$) inside a quiescent prominence is nearly equal to the gas pressure outside it. The mechanical support of these arches is provided by long lasting magnetic field formations. These magnetic fields prevent also the quick dispersion of the dense plasma of a quiescent prominence into the surrounding corona. The temperature of an active prominence is about 30,000°K and its spectrum is substantially different from the spectrum of a quiescent prominence. Prominences are classified into many categories according to their shapes, spectra, and motions of their matter. *Eruptive prominences* and *surge prominences* (surges) have been observed shooting matter into the corona at speeds of nearly 1,000 km/sec and have been recorded in some excellent sequences of time-lapse photography.

4.6 Radio and X-Ray Bursts from the Sun

In the specific meaning of the term, a solar flare is the sudden brightening of a small region of the solar disc. More generally, however, it is an explosive event in the atmosphere of the sun which produces not only optical effects but also intense radio, x-ray, and corpuscular radiation. While the optical flux from the sun never varies by more than 1%, the enhancement in the x-ray region and in the radio domain during a flare event can exceed the respective flux from the quiet sun by several orders of magnitude. For this reason all modern studies of solar activity are now concentrating in these two extreme regions of the solar spectrum. X-ray bursts from the sun can be detected either directly with special instruments flown on rockets and satellites (see Section 7.2), or indirectly from the ground by the effects which they produce in the terrestrial ionosphere (see Section 6.5). Solar radio bursts are observed directly from the ground through the radio window of the terrestrial atmosphere.

When a flare event is triggered in an active region on the sun, it produces a sharp burst of energetic electrons and protons which stream with velocities of the order of $V \sim 10^{10}$ cm/sec both outwards toward the corona and inwards toward the chromosphere (Figure 4.6-I). The shooting of these particles is most probably the result of the pinching of the magnetic field near the neutral point and these bursts consist often of 2–3 spurts, about ten seconds apart, which remind one of the repeated pinching effect observed in the laboratory. Certain flare events produce two or more such particle outbursts in a time interval of several minutes, which is an indication that the magnetic field of the active region had two or more neutral points where an instability could develop.

The downward stream of the relativistic electrons produces immediately a burst of hard x-rays through collisions either with ions in the corona or with neutral hydrogen atoms in the chromosphere. Since this radiation is produced by a rather well collimated beam of energetic particles, we are dealing with a non-thermal process. For this reason this emission mechanism is called *non-thermal bremsstrahlung*. The reason that the same effect is produced in encounters with both ions and hydrogen atoms is because, in order to produce photons of $\lambda \sim 1$ Å, the high energy electrons must come much closer to the nuclei of the hydrogen atoms than the radial distance of their orbiting electrons.

This can be seen by equating the energy hc/λ of the emitted photon with the Coulomb potential energy e^2/r of the electron-proton encounter,

$$r = \frac{e^2\lambda}{hc} \simeq 10^{-3}\lambda \qquad (4.6\text{-}1)$$

Thus for $\lambda \sim 1$ Å we find that $r \sim 10^{-11}$ cm, which is much smaller than the $\sim 10^{-8}$ Bohr radius of the hydrogen atom, and therefore the presence of an orbital electron is of little importance. The duration of a non-thermal x-ray burst, which is also called a *class-II* x-ray burst, is of the order of a few minutes. Occasionally, however, we might have two or more such bursts over a period of several minutes that are all associated, as we mentioned above, with the same general flare event. The spectrum of the class-II bursts peaks below 1 Å (Figure 7.2-I) and for the bigger events it can be detected all the way into the γ-ray region ($\lambda < 0.02$ Å).

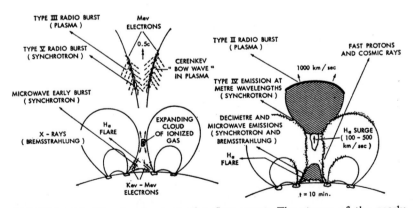

FIGURE 4.6-I   The development of a flare event. The stream of the nearly relativistic electrons (left) and the large plasma cloud (right) which are ejected from the flare region, produce many different types of x-ray, microwave, and meter radio bursts (Hess and Mead, 1968)

The stream of these high energy electrons is thermalized quickly in the denser layers below the center of the flare, which accounts for the short duration of the non-thermal x-ray bursts. In exchange, however, these layers are heated up to a temperature of 10^7–$10^{8\circ}$K and start emitting x-ray radiation through *bound-bound* (line emission), *bound-free* (recombination emission) and *free-free* (thermal bremsstrahlung) transitions. This constitutes a thermal x-ray burst, which is also called a *class-I* x-ray burst. The spectrum of the class-I bursts peaks in the 1–10 Å range and therefore it is softer than the class-II but harder than the x-ray spectrum emitted by the quiet corona.

The rise time of the intensity curve of the thermal x-ray bursts coincides, and is of approximately equal duration, with the non-thermal x-ray burst. The duration of the thermal x-ray bursts increases at longer wavelengths, but typically ranges from 10 minutes to 1 hour. The relatively long duration of the class-I bursts represents the prolonged cooling period of the layers that were heated up by the initial stream of energetic particles. Consequently, when a class-II burst has several peaks and valleys around 0.3 Å, due to two or more independent particle bursts, the intensity curve of the corresponding class-I burst around 10 Å shows only a step-like structure (Figure 4.6-III) because of the slow decaying time of the thermal x-ray burst.

Concurrently with the non-thermal x-ray burst we usually receive also a strong radio burst in the microwave (millimeter and centimeter) region of the spectrum. In general there is very good agreement in the timing of the two bursts and in the shapes of their intensity curves (Anderson and Winckler, 1962). This suggests that the *microwave burst* is produced by the same high energy electrons which produce the non-thermal x-ray burst, probably through synchrotron emission in the strong magnetic fields in the region near and below the center of the flare explosion. The microwave burst lasts also only a few minutes and occurs usually during the rapid rise of the optical flare to its peak intensity. This brief microwave burst is often followed by a *post-burst increase* of the solar flux in the centimeter region which most likely is produced by thermal bremsstrahlung. The post-burst increase, like the optical flare and the thermal x-ray burst lasts for a period of the order of one hour.

Besides the downward burst of high energy electrons, there is usually also an upward burst of energetic particles which speed through the solar corona exciting plasma waves. These compression waves of the coronal plasma use part of their energy to generate radio waves at the local plasma frequency f_N (2.3-19). As these energetic particles move with high velocities ($V \sim 10^{10}$ cm/sec) outwards through layers of continuously lower electron density, the peak frequency of the radio burst $(f = f_N = A N_e^{1/2})$ drifts rapidly toward lower values. These solar radio bursts are called *type-III* or *fast drift radio burst* (Figure 4.6-II). Type-III bursts are usually observed in the meter region ($300 < f < 30$ MHz) through which they drift in approximately 10–20 seconds. The drifting rate of the type-III bursts varies with frequency being of the order of 30 MHz/sec around 100 MHz and about 1 MHz/sec around 10 MHz.

Satellite observations above the maximum of the terrestrial ionosphere, which imposes a frequency cut-off around 10 MHz, have shown that the frequency drifts of certain type-III bursts continue all the way into the hectometer region ($3 < f < 0.3$ MHz). The duration of a type-III burst, as measured at a given frequency, increases as the frequency decreases but is typically of the order of a few seconds. Frequently each burst consists of 2 or 3 individual peaks separated by about 10 seconds from each other which probably is due to the repeated pinching effect at the same neutral point. In some of the larger flare events we might also observe a sequence of two or more of these closely bunched bursts over a period of several minutes (Figure 4.6-III) which probably is due to succesive outbursts of energetic particles from different neutral points in the general region of the flare.

Occasionally the upward burst of the energetic electrons cannot escape the strong magnetic field above the active region and after reaching a maximum height in the corona it is forced to fall back again toward the chromosphere. Under these circumstances, the peak frequency of the radio burst first decreases to a minimum and then drifts back toward higher frequencies. These type-III bursts are called *U-type* bursts from the shape of their dynamic spectra which resembles an inverted letter *U*.

Type-III bursts are frequently followed by a *type-V* radio burst (Figure 4.6-II) which is simply a continuum of radio noise over a wide band of meter wavelengths without a specific peak frequency. Type-V bursts last 1–3 minutes and their emission mechanism is believed to be synchrotron radiation. Some of the energetic particles that produce the type-III bursts are probably deflected from the main stream, and as they speed through the regular magnetic field of the sun (the field of the coronal streamers) they emit synchrotron radiation at meter wavelengths (Figure 4.6-I).

Following the initial explosion of the flare, an expanding cloud of ionized gas is formed which moves outwards with a velocity of about 1,000 km/sec. Velocities of this order are much higher ($\sim$ Mach 10) than the speed of sound in the solar corona and result in the formation of a shock wave which advances in front of the plasma cloud (Figure 4.6-I). As the shock wave propagates through the solar corona, it excites longitudinal plasma oscillations (plasma waves) which in turn produce radio waves at the local plasma frequency. The resulting radio bursts display again a frequency drift which, however, is much slower (~ 1 MHz/sec, at ~ 100 MHz) than the frequency drift of the type-III bursts. The reason of course is that the velocity of the shock front is about 100 times slower than the velocity of the energetic particles which produce the type-III bursts. These bursts are called *type-II* or *slow drift radio bursts*.

The type-II bursts are often accompanied by a second harmonic, i.e., a burst at twice the fundamental frequency (Figure 4.6-II). Careful observations have not revealed the presence of any higher harmonics which suggests that the appearance of the second harmonic is not due to non-linear effects. Ginzburg and Zheleznyakov (1958) have suggested that the *coupling mechanism* (the mechanism through which the energy of the plasma waves is transformed into electromagnetic waves) at the fundamental frequency is scattering of electron plasma waves from ionic charge fluctuations, while the second harmonic at twice the local plasma frequency, is produced by the scattering of electron plasma waves from electron density fluctuations.

Type-II bursts first appear around 100–300 MHz, which is the local plasma frequency at the height where the shock front develops enough power to start exciting plasma oscillations. Their frequency drift to about 30 MHz lasts approximately 10–30 minutes and the duration of a type-II burst at a single frequency is of the order of a few minutes. It should be pointed out that the drift rates of both type-II and type-III bursts vary somewhat from case to case. Their respective ranges, however, come no way near each other so that there is no doubt that type-II and type-III radio bursts are excited by different sources.

The plasma cloud, which produced a type-II burst, ultimately leaves the sun and continues to propagate in the interplanetary space with a

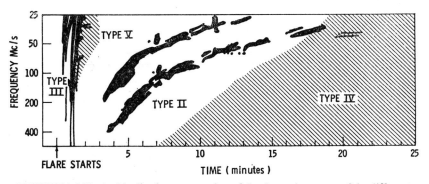

FIGURE 4.6-II   An idealized representation of the dynamic spectra of the different types of solar radio bursts in the meter wavelength range (Hess and Mead, 1968)

velocity of ~1,000 km/sec. If the geometry is right, this cloud will encounter the magnetosphere of the earth in about a day and a half,

$$t = \frac{L}{V} = \frac{1.5 \times 10^{13}}{10^8} = 1.5 \times 10^5 \text{ sec} \sim 1.5 \text{ days} \qquad (4.6\text{-}2)$$

causing geomagnetic storms, auroras, etc. Type-II bursts are intense but rather infrequent events. Even near the peak of the solar cycle they occur only once every other day while we might have several hundred of the type-III bursts per day. Sweep frequency or multi-channel receivers can show on an appropriate display (e.g. an oscilloscope) the *dynamic spectra* of type-II and type-III bursts. From these frequency versus time curves, one can in principle derive the electron density profile of the solar corona assuming known values for the velocities of the exciters (see 7.4-IV).

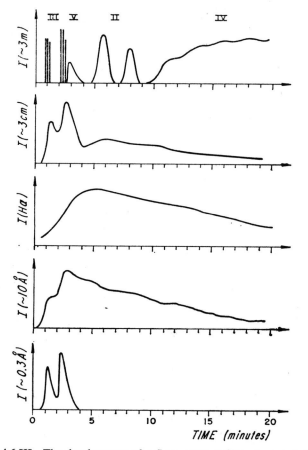

FIGURE 4.6-III   The development of a flare event at fiive characteristic wavelengths of the solar spectrum

Type-II bursts are often followed by a persistent, nearly featureless radio emission over a broad band of wavelengths in the meter range. This smooth continuum is called a *type-IV* burst (Figure 4.6-II) and can last from less than one hour to a few days. Type-IV bursts are most probably synchrotron radiation emitted by some energetic flare electrons which become trapped in the magnetic field of the active region.

Figure 4.6-III shows the development of a typical flare event in five representative wavelengths of the solar spectrum. Besides the development of the optical flare in the Hα-line of hydrogen, Figure 4.6-III shows also the intensity changes in four other spectral regions, namely in the domains of: the meter (1–10 m) radio burst, the microwave (0.5–10 cm) radio bursts, the thermal (1–20 Å) x-ray bursts, and the non-thermal (0.1–1 Å) x-ray bursts. By studying a flare event over such a wide range of wavelengths, we enhance tremendously our opportunities to comprehend fully the physics of this intricate and far reaching phenomenon.

4.7   The Development of an Active Region on the Sun

Now that we have studied all the different manifestations of the solar activity, we must try to tie them together into a chronological sequence of events. For this reason we will summarize in this section the life history of an active region of the sun. Note that according to an established convention the sun rotates from the east limb to the west limb. As seen from Figure 4.7-I, a terrestrial observer of the sun from a position on the noon meridian has west to his right both for the earth and the sun.

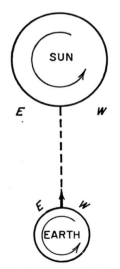

FIGURE 4.7-I   The convention for East and West on the earth and on the sun

The first sign usually is the appearance of a small bright speck which is the beginning of the facular region. If this region is located near the limb it might be visible in white light, but at any rate it will always be visible in the Hα-line of hydrogen or the K-line of calcium (plages). The magnetic field of the region increases rapidly to 100 Gauss or more and a few hours later the first dark pores appear.

During the following day the preceding sunspot is formed from the pores at the west end of the facular region and in another day or two it develops a penumbra. A spot of opposite magnetic polarity, called the following spot, appears and forms a penumbra at the east end of the facular region with a delay of one to two days. Several other smaller spots appear between the two main ones forming a sunspot group with 10–20 spots.

Meanwhile the plage area increases in brightness and size, to a dimension of about 100,000 km. The first flare might occur on the fourth day, but the flare activity reaches a maximum around the tenth day when the sunspots reach also their maximum development. Thus the second week is actually the week of maximum flare activity.

During the following two weeks the sunspots disappear. The following spot breaks up first into smaller ones and then vanishes together with the other smaller spots. The preceding sunspot will probably be the only spot still visible at the end of the first month. During the third and fourth week flares are rare, but the plage area continues to increase in size while the magnetic field of the active region reaches its maximum flux during this period. At this stage we observe also the appearance of a thin, long filament which forms an angle of approximately 40° to the meridian and points toward the still surviving preceding spot. If the active region happens to be near the limb, the filament appears as a prominence arching beyond the disc of the sun.

In the second month the remaining spot disappears, the brightness of the facular region decreases, and the distribution of the magnetic field in the active area becomes very irregular. The filament, however, grows steadily in size reaching a length of about 100,000 km. Now, because of the differential rotation of the sun, it is orientated more along the east-west direction and divides the active region into roughly two halves.

During the third and fourth month we have no more spots, the facular region disappears, and the magnetic field declines in strength but it is still higher than normal. The filament reaches its maximum length and becomes almost parallel to the equator.

The magnetic field might still be detectable after the sixth month. During this period it continues to spread out over an even larger area while it steadily weakens in strength until it finally merges with the general magnetic field of the sun. The filament might also remain visible up to the sixth month, decreasing in size, changing in structure, and finally migrating toward the polar regions. Finally at approximately the same time all the last remnants of an active region disappear, but this might be more than half a year from the appearance of the first sunspots.

It should be mentioned here that new sunspots have the tendency to appear in the vicinity of decaying active regions, so it is not uncommon that an active region will have several revivals. As a matter of fact it has been observed that certain general areas of the sun remain active for periods of a year or more (Sawyer, 1968). This explains a general periodicity of 27 days (the rotation of the sun) in solar activity, which otherwise is difficult to understand on the basis of isolated active regions where flare activity usually lasts for less than a month.

In the preceding paragraphs we have summarized the life history of a typical center of solar activity. In Figure 4.6-II we showed typical dynamic spectra of meter solar radio bursts and in Figure 4.6-III we depicted the development of a typical flare event in five different spectral regions. It is obvious that the word *typical* is used quite frequently and for this reason it would be useful to try to clarify its meaning. In the study of complicated natural phenomena, such as the solar activity, one can hardly expect to find two events that are exactly the same. For this reason it is necessary to distill a model event from a large number of actual events. In this model we must incorporate all the basic features of the phenomenon we are trying to study, but we must leave out all the intrinsic idiosyncrasies of the individual events. The result of course is that essentially none of the actual events look exactly like the model event and quite often they miss the one or the other general feature we have incorporated into the model event. Thus we can observe type-III bursts without type-II bursts, x-ray bursts without radio bursts, optical flares without x-ray bursts, etc. To some extent this is similar to the concept of the All-American boy or girl. These supposedly "typical" samples incorporate most of the good general features that one encounters in American boys and girls but very few of their individual problems. At the end, these All-American prototypes do not look very much like any of the boys and girls you know.

Nevertheless an idealized event has great usefulness not only for educational purposes, but also for the theoretician who is trying to understand the basic physics of this phenomenon. Retaining the essential features and leaving out the complicating details makes it possible to build a simple physical model that displays the basic processes of the phenomenon we are studying. Such simple models make possible the mathematical formulation of the phenomenon from which one can draw theoretical predictions and thus check the validity of the model against experimental observations. Special features or modifications can be incorporated later into the basic model in order to explain the occasional appearance of certain less important or less general effects. Trying to account for all the complexities of the phenomenon from the beginning by starting with a very detailed model can lead one to a totally unweildy mathematical problem. Furthermore, there is a good chance that the mathematical complexity of the problem will sidetrack one from the actual physics of the phenomenon, and as people say, one might miss the forest for the trees.

It is therefore very essential to strip the different events of their individual details and from a large number of cases to distill a simple model, a *typical event* as it is frequently called, which reflects accurately the basic physical processes involved and which is simple enough to make possible a reasonable mathematical formulation of the phenomenon.

4.8 Bibliography

A. Books for Further Studies

1. *Sunspots*, R. J. Bray and R. E. Loughhead, Barnes and Noble, Inc., New York, N. Y., 1964.
2. *Solar Flares*, H. J. Smith and E. P. Smith, The MacMillan Co., New York, N. Y., 1963.
3. *The Solar Granulation*, R. J. Bray and R. E. Loughhead, Barnes and Noble, Inc., New York, N. Y., 1967.
4. *A Guide to the Solar Corona*, D. E. Billings, Academic Press, New York, Y.,N. 1966.
5. *The Solar Atmosphere*, H. Zirin, Blaisdell Publ. Co., Waltham, Mass., 1966.
6. *Solar Activity*, Einar Tandberg-Hanssen, Blaisdel Publ. Co., Waltham, Mass., 1966.
7. *Solar Radio Astronomy*, M. P. Kundu, John Wiley and Sons, New York, N. Y., 1965.
8. *The Solar Spectrum*, ed. by C de Jager, D. Reidel Publ. Co., Dordrecht, Holland, 1965.
9. *The Sun*, ed. by G. P. Kuiper, The University of Chicago Press, Chicago, Ill., 1953.
10. *The Sun and the Stars*, J. C. Brandt, McGraw Hill, New York, N. Y., 1966.
11. *Space Physics*, R. S. White, Gordon and Breach, New York, N. Y., 1970.
12. *Solar System Astrophysics*, J. C. Brandt and P. W. Hodge, McGraw Hill, New York, N. Y., 1964.
13. *Astrophysics and Space Science*, A. J. McMahon, Prentice Hall, Englewood Cliffs, N. J., 1965.
14. *The Physics of Solar Flares*, ed. by W. N. Ness, NASA SP-50, U. S. Government Printing Office, Washington, D. C., 1964.
15. *Introduction to Space Science*, ed. by W. H. Ness and G. D. Mead, Gordon and Breach, New York, N. Y., 1968.
16. *Research in Geophysics*, ed. by H. Odishaw, M. I. T. Press, Cambridge, Mass., 1964.
17. *Space Physics*, ed. by D. P. LeGalley and A. Rosen, John Wiley and Sons, New York, N. Y., 1964.
18. *Solar Physics*, ed. by John Xanthakis, John Wiley and Sons, New York, N. Y., 1967.

19. *Introduction to Solar Terrestrial Relations*, ed. by J. Ortner and H. Maseland, Gordon and Breach, New York, N. Y., 1965.

20. *Mass Motions in Solar Flares and Related Phenomena*, ed. by Yngve Ohman, John Wiley and Sons, New York, N. Y., 1968.

B.   Articles in Scientific Journals

Anderson, K. A. and J. R. Winckler, Solar flare x-ray burst on September 28, 1961, *J. Geophys. Res.*, **67**, 4103, 1962.

Babcock, H. W., The topology of the sun's magnetic field and the 22-year cycle, *Astrophys. J.*, **133**, 572, 1961.

Friedman, H., Ultraviolet and x-rays from the sun, *Ann. Rev. of Astron. and Astrophys.*, **1**, 59, 1963.

Ginzburg, V. L. and V. V. Zhelezniakov, On the possible mechanisms of sporadic solar radio emission, *Soviet Astronomy A. J.*, **2**, 653, 1958.

Gnevyshev, M. N., On the 11-year cycle of solar activity, *Solar Phys.*, **1**, 107, 1967.

Gold, T. and F. Hoyle, On the origin of solar flares, *Mo. Not. Roy. Astron. Soc.*, **120**, 84, 1960.

Leighton, R. B., A magneto-kinematic model of the solar cycle, *Astrophys. J.*, **156**, 1, 1969.

Sawyer, C., Statistics of solar active regions, *Ann. Rev. Astron. Astrophys.*, **6**, 115, 1968.

Severny, A. B., The appearance of flares in neutral points of the solar magnetic field and the pinch effect, *Izvest. Krimskoi Astrofiz. Obs.*, **20**, 22, 1958.

Sweet, P. A., The neutral point theory of solar flares, *Proc. I. A. U.*, Symp. No. 6, p. 123, 1958.

Xanthakis, J., On the relation between the indices of solar activity in the photosphere and the corona, *Solar Phys.*, **10**, 168, 1969.

CHAPTER 5

THE INTERPLANETARY SPACE

5.1 Introduction

The casual concept of free space is that of total emptiness and darkness. This, however, is not a correct picture for the interplanetary space because the space around the sun and the planets contains dust, charged particles, and magnetic fields, which all together constitute the *interplanetary medium*. The interplanetary space is of course also permeated by the intense electromagnetic radiation from the sun which is often scattered from the electrons and the dust particles of the interplanetary space.

Actually it would have not been too difficult to guess the nature of the interplanetary medium from simple observations and basic physical principles, but for a long time our thinking was hampered by certain preconceived ideas. It is not very difficult for example to predict that a continuous evaporation of solar plasma must take place when a star with a hot corona is surrounded by cold, empty space. For a long time, however, people thought that the strong gravitational field of the sun would not allow the escape of the evaporating plasma into the interplanetary space. It turns out that this is true only up to a distance of a few solar radii, because further out the thermal energy of the plasma overtakes the gravitational energy of the sun and the corona starts to expand supersonically into the interplanetary space. This continuous outflux of coronal gas into the interplanetary space is called the *solar wind*.

From Maxwell's equations one can also see that a plasma of very high temperature and therefore of very high conductivity sweeps with it as it moves all external magnetic fields. The reason is that the magnetic fields remain *frozen in* because the total magnetic flux through the surface of any changing volume element of a highly conducting plasma remains constant ($d\phi/dt = 0$). Since the sun possesses a magnetic field, we should have expected to find also magnetic fields in the interplanetary space, transported there by the solar wind.

Finally, from the comets that are often seen to disintegrate in the vicinity of the sun, and the meteors that bombard continuously the earth, it should have been quite obvious that the interplanetary space must contain also a great variety of dust particles. In conclusion, a picture of total emptiness for the interplanetary space was by no means justified and today with in situ observations we know that it is not correct. Of course the plasma density, the intensity of the magnetic field, and the density of the dust particles of the interplanetary medium are extremely low with terrestrial standards, but the fact is that they are there and make the interplanetary space a very interesting region to study.

5.2   Characteristic Parameters of Fully Ionized Plasmas

Fully ionized plasmas, such as the solar wind, are frequently encountered both in space physics and in space astronomy. For this reason it might be a good opportunity to precede the section on the solar wind with a brief review of the different parameters that characterize a fully ionized plasma. Some of the important length parameters are:

The *Debye length* l_D. This is the distance at which the electromagnetic effects of a charged particle cease to be important due to the screening action of all the other charged particles around it. The Debye length is given by the expression,

$$l_D = \frac{V}{3^{1/2}\omega_N} = \left(\frac{mV^2}{12\pi e^2 N}\right)^{1/2} = \left(\frac{kT}{4\pi e^2 N}\right)^{1/2}$$

$$= 6.9\left(\frac{T}{N}\right)^{1/2} \text{ cm} \tag{5.2-1}$$

where T is in °K and N in el/cm³. Note the dependence on the plasma frequency $\omega_N = 2\pi f_N$ (2.3-12), which is probably the most important parameter of a fully ionized plasma.

The *scale height* H_c. This is the distance over which the plasma density of the corona changes by a factor of $1/e$. For an isothermal corona in hydrostatic equilibrium,

$$H_c = \frac{kT}{mg} = \frac{kTr^2}{mGM_0} = 6 \times 10^3 T\left(\frac{r}{R_0}\right)^2 \text{ cm} \tag{5.2-2}$$

where in (5.2-2) we have included also the dependence of the acceleration of gravity g on the radial distance r from the sun in solar radii R_0.

The *gyro-radius* r_H. This is the radius of gyration of a charged particle moving in a magnetic field,

$$r_H = \frac{V}{\omega_H} = \frac{\left(\frac{3kT}{m}\right)^{1/2}}{\left(\frac{eH}{mc}\right)} = \frac{(3kTmc^2)^{1/2}}{eH} \tag{5.2-3}$$

Note that the gyro-frequency $f_H = \omega_H/2\pi$, is also a very important parameter. Due to the difference in mass, the electrons have a much smaller gyro-radius than the protons. For H in Gauss and T in °K, the electron gyro-radius is,

$$r_{He} = \frac{0.038T^{1/2}}{H} \text{ cm} \qquad (5.2\text{-}4)$$

and the proton gyro-radius is,

$$r_{Hp} = \frac{1.6T^{1/2}}{H} \text{ cm} \qquad (5.2\text{-}5)$$

The *mean free path* λ_e, which is the distance a free electron travels between succesive collisions with ions,

$$\lambda_e = \frac{V_e}{\nu_i} = \frac{V_e}{\sigma_i N V_e} = \frac{1}{\sigma_i N} \sim 10^4 \frac{T^2}{N} \text{ cm} \qquad (5.2\text{-}6)$$

where ν_i (2.4-3) is the collision frequency of electrons with ions and σ_i (2.4-2) the corresponding cross-section which is proportional to T^{-2}. In the numerical result of (5.2-6) N is the plasma density in el/cm³.

Table 5.2-I gives the orders of magnitude of all these characteristic lengths (l_D, H_c, r_{Hp}, r_{He}, and λ_e) first in the corona and then near the orbit of the earth. In the corona the basic parameters are: $T \sim 10^6$°K, $N \sim 10^8$ el/cm³ and H $\sim$ 1 Gauss, while at 1 A.U. the corresponding values are, $T \sim 10^5$°K, $N \sim 10$ el/cm³ and $H \sim 5 \times 10^{-5}$ Gauss.

TABLE 5.2-I

	l_D	H_C	r_{Hp}	r_{He}	λ_e (in cm)
Corona	1	10^{10}	10^3	10^1	10^8
1A.U.	10^3	10^{13}	10^7	10^5	10^{13}

As seen from Table 5.2-I, the Debye length is always much smaller than any of the other characteristic lengths and this fact guarantees that the basic principles of plasma physics apply to the interplanetary medium. This table shows that the mean free path in the interplanetary space is extremely large, becoming of the order of 1 A.U. at the distance of the earth. This means that we are dealing essentially with a collisionless plasma which, it would seem, could not be treated as a fluid. It is also seen, however, that the radii of gyration are much smaller than λ_e and H_c and thus the magnetic field holds the interplanetary plasma together and allows it to be treated as a continuous fluid. It is this factor that justifies the application of the principles of hydrodynamics in the flow of the solar wind.

The thermal conductivity of a fully ionized gas is proportional to the temperature of the gas to the 5/2 power. As a result the thermal conductivity of the corona with a temperature $T \sim 10^6°K$ is extremely high and heat diffuses through it very rapidly. This explains why the temperature of the corona and of the interplanetary plasma change very slowly with distance from the sun. One can obtain the temperature dependence of the thermal conductivity using very simple physical principles. Let dT/dx be the temperature gradient of a gas in the x-direction and N the particle density of the gas. The thermal conductivity K is defined by the equation,

$$\frac{dQ}{dt} = KA \frac{dT}{dx} \tag{5.2-7}$$

where dQ is the amount of energy crossing in time dt an area A which is normal to the x-axis. The rate R_1 at which particles move from region 1 to region 2 through the area A (Figure 5.2-I) will be proportional to $N_1 VA$,

$$R_1 \propto N_1 VA \tag{5.2-8}$$

and since each particle carries an amount of energy proportional to kT, the energy dQ_1 crossing the area A in time dt from region 1 to region 2 will be proportional to,

$$dQ_1 \propto R_1 kT_1 \, dt \propto AN_1 VkT_1 \, dt \tag{5.2-9}$$

The particles which have a chance to cross from the one region to the other cannot be more than a mean free path λ away from A. This determines the volume of the two regions, and from the given temperature

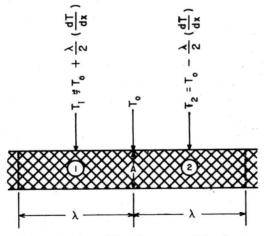

FIGURE 5.2-I   The conduction of heat by gas particles due to a temperature gradient in the gas

gradient dT/dx, it also specifies their respective average temperatures T_1 and T_2. As seen from Figure 5.2-I, the temperature difference $T_1 - T_2$ is given by the relation,

$$(T_2 - T_1) = \lambda \left(\frac{dT}{dx} \right) \tag{5.2-10}$$

Assuming now for simplicity that $N_1 = N_2 = N$, we find that the net energy flux through the area A is proportional to,

$$\frac{dQ}{dt} = \frac{dQ_1}{dt} - \frac{dQ_2}{dt} \propto ANVk(T_1 - T_2) \propto ANVk\lambda \frac{dT}{dx} \tag{5.2-11}$$

Finally by comparing (5.2-11) with (5.2-7) we find that the thermal conductivity K is proportional to,

$$K \propto kNV\lambda \propto \frac{kNV^2}{v} \propto \frac{kNV^2}{\sigma NV} \propto \frac{kV}{\sigma} \tag{5.2-12}$$

where σ is the collisional cross-section. For a neutral gas σ is independent of the temperature ($\sigma \sim 10^{-15}$ cm^2), and since V is proportional to $(kT/m)^{1/2}$ we find that K is proportional to $T^{1/2}$. For a fully ionized plasma, however, as seen from (2.4-2) σ is proportional to $(e^2/kT)^2$. Thus by introducing the temperature dependence of V and σ in (5.2-12) we find that for a fully ionized plasma K is proportional to $T^{5/2}$. The actual expression is,

$$K \propto \frac{k^{7/2}T^{5/2}}{m^{1/2}e^4} \tag{5.2-13}$$

The constant of proportionality for (5.2-13) can be evaluated accurately only through a much more elaborate derivation but still turns out to be a number of the order of unity.

From (5.2-13) it is obvious that the thermal conductivity of the electron gas is $(m_e/m_p)^{1/2} \sim 50$ times higher than the thermal conductivity of the proton gas. One could therefore expect the electron gas at large distances from the sun to be hotter than the proton gas. This is found to be true and at 1 A.U. the most probable proton and electron temperatures are respectively $T_p \simeq 5 \times 10^{4}$°K and $T_e \simeq 1.5 \times 10^{5}$°K, i.e., $T_e/T_p \simeq 3$. One should remember, however, that the temperature of a collisionless plasma is a somewhat vague concept and the temperatures quoted, which were obtained with satellites, were reduced from the distribution of the random velocities of the electrons and the protons, relative to the mean bulk velocity of the solar wind, assuming a Maxwellian distribution of particle velocities. These measurements have also shown that the ions possess a field aligned temperature anisotropy, i.e. observations along the magnetic field yield a temperature which is approximately twice the temperature obtained at right angles to the magnetic field.

When there is little interaction between protons and electrons one must use a 2-fluid model for the solar wind, which complicates things considerably. In general, however, we can assume that protons and electrons move together in order to maintain the neutrality of the plasma. This is called *ambipolar diffusion* and the plasma can be treated as a single fluid with a thermal conductivity equal to twice the thermal conductivity of the proton gas. Using (5.2-13) with the correct constant of proportionality and the fact that $K = 2K_p$, we find that the thermal conductivity of a single fluid plasma with $T = T_e = T_p$ is,

$$K \simeq 10^{-6}T^{5/2} \text{ erg sec}^{-1} \text{ cm}^{-1} \text{ deg}^{-1} \qquad (5.2\text{-}14)$$

The *electrical conductivity* σ, (not to be confused with the cross-section we have used for which the same symbol) is another important plasma parameter and is defined by the relation,

$$J = \sigma E \qquad (5.2\text{-}15)$$

where J is the current, i.e. the flux of charged particles, and E is the electric field that generates the current. In a fully ionized plasma σ is proportional to $T^{3/2}$ and is given by the expression,

$$\sigma = \frac{e^2 N}{mv_i} \propto \frac{k^{3/2}T^{3/2}}{e^2 m^{1/2}} \qquad (5.2\text{-}16)$$

The dependence of v_i on T was obtained in (2.4-3). Equation (5.2-16) can be derived also with the use of only very basic physics. Let us apply for example a constant electric field on a fully ionized plasma. If the field is not very strong, only the electrons will tend to move under the force of the electric field, while the heavier ions can be considered as stationary. Under these conditions the equation of motion of the electrons is,

$$m \frac{dV}{dt} = eE - mVv_i \qquad (5.2\text{-}17)$$

where mNv_i is the dragging force due to the collisions of the electrons with the ions (2.4-6). The velocity V of the electrons relatively to the stationary ions will continue to increase $(dV/dt > 0)$ until the electric force eE is balanced by the frictional force mVv_i. At this point the velocity will stop increasing $(dV/dt = 0)$ and we will reach a steady state with,

$$V = \frac{eE}{mv_i} \qquad (5.2\text{-}18)$$

The current J, produced by the motion of the electrons is,

$$J = NeV \tag{5.2-19}$$

and from (5.2-18) and (5.2-19) we obtain,

$$J = \left(\frac{Ne^2}{mv_i}\right) E \tag{5.2-20}$$

Finally, by comparing (5.2-20) with (5.2-15) we deduce the expression for σ (5.2-16).

5.3 Hydrodynamic Equations in the Solar Corona

The solar wind, as we have seen, can be considered as a fluid in motion. For this reason in our treatment of the outer corona we must replace the hydrostatic equation (the gas pressure balancing the gravitational force) with the hydrodynamic equation. This simply means that in addition to gravity and pressure we must also include a force due to the forward motion of the fluid. This third force is actually the transport of momentum by the particles and is equal to,

$$mV\left(\frac{dV}{dr}\right) \tag{5.3-1}$$

which, as one can easily see, has the dimensions of force (MLT^{-2}). To derive this expression we must start with the equations for the conservation of energy and momentum of the moving fluid. Let A be the cross-section of a column in the direction of motion of this fluid (Figure 5.3-I), which of course is the radial direction in the solar corona. Let N be the particle density of the fluid, m the average mass of the particles, and V their velocity along the axis of the column. The particle flux through the cross-section A of the cylinder of Figure 5.3-I will be equal to NVA. The total number

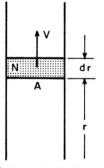

FIGURE 5.3-I   A column of plasma in hydrodynamic equilibrium in the solar corona

of particles, on the other hand, in a volume element $dv = A \, dr$ is $N \, dv$. Any change in the flux of particles in the cylinder will produce a change in the number of particles in the volume element dv. Since the particles have a mass m, we can write the above statement in the form of an equation for the conservation of mass,

$$d(mNVA) = -\frac{d}{dt}(mN \, dv) \qquad (5.3\text{-}2)$$

The L.H.S. of (5.3-2) represents the difference of the mass leaving minus the mass entering the volume element dv per unit time, while the R.H.S. of (5.3-2) represents the mass change per unit time in the volume element dv. The minus sign indicates that when the flux out of the box increases, the mass in the box decreases. Using $A \, dr$ for dv in (5.3-2), and assuming that A and dr are not functions of time, we get,

$$\frac{d}{dr}(mNV) = -\frac{d}{dt}(mN) \qquad (5.3\text{-}3)$$

Assuming now that V remains constant ($dV/dt = 0$), we can write,

$$\frac{d}{dt}(mNV) = V\frac{d}{dt}(mN) + mN\frac{dV}{dt} = V\frac{d}{dt}(mN) \qquad (5.3\text{-}4)$$

and by multiplying (5.3-3) by V and using (5.3-4) we get the final equation for the conservation of mass,

$$\frac{d}{dt}(mNV) = -V\frac{d}{dr}(mNV) \qquad (5.3\text{-}5)$$

The momentum mV of all the particles in the volume element dv is $(mV) N \, dv$, and the flux of the momentum is $(mN) NAV$. A change in this flux must be equal to the change per unit time of the momentum in dv. But the time change of momentum is a force, and therefore we must include also in the equation any other forces that act on the mass in the volume element dv. Since we are interested in a fluid moving radially out in the solar corona, we must include here the pressure gradient in the corona and the gravitational force of the sun. Thus the equation for the conservation of momentum takes the form,

$$d\{(mV) NAV\} = -\frac{d}{dt}\{(mV) NAdr\} - (dP) A - g(mNAdr) \qquad (5.3\text{-}6)$$

where dP is the pressure difference above and below the volume element, and g is the acceleration of gravity of the sun,

$$g = \frac{GM_0}{r^2} \qquad (5.3\text{-}7)$$

Since A and dr are independent of time, we can write (5.3-6) as follows,

$$\frac{d}{dr}\{(mNV)\,V\} = -\frac{d}{dt}(mNV) - \frac{dP}{dr} - \frac{GM_0 mN}{r^2} \qquad (5.3\text{-}8)$$

or,

$$V\frac{d}{dr}(mNV) + mNV\left(\frac{dV}{dr}\right) = -\frac{d}{dt}(mNV) - \frac{dP}{dr} - \frac{GM_0 mN}{r^2} \qquad (5.3\text{-}9)$$

we can now use (5.3-5) to cancel the first terms on the L.H.S. and the R.H.S. of (5.3-9). The remaining terms give the hydrodynamic equation for the solar corona.

$$\frac{GM_0 mN}{r^2} = -\frac{dP}{dr} - (mNV)\frac{dV}{dr} \qquad (5.3\text{-}10)$$

The only difference between the hydrodynamic equation and the hydrostatic equation which we had used for the terrestrial atmosphere, is the last term of (5.3-10), which as we mentioned in the beginning of this section, represents the force due to the motion of the fluid.

For a plasma where $N = N_e = N_i$, the total pressure (gas pressure) is the sum of the ion gas pressure and the electron gas pressure. Therefore, if $T = T_e = T_i$, we have,

$$P = 2NkT \qquad (5.3\text{-}11)$$

Introducing (5.3-11) in (5.3-10) and dividing all terms by mN we obtain,

$$\frac{1}{mN}\frac{d}{dr}(2NkT) + V\frac{dV}{dr} + \frac{GM_0}{r^2} = 0 \qquad (5.3\text{-}12)$$

It should be clarified that for a fully ionized plasma m in (5.3-12) represents the mass of the ions, because both the mass and the momentum of the electrons are negligible compared to the ions.

From the spherical symmetry of the problem, and the fact that particles flow continuously out without accumulating in any particular layer of the corona, it follows that the total flux of particles through any spherical shell around the sun is constant,

$$4\pi r^2(NV) = 4\pi R_0^2(N_0 V_0) = \text{constant} \qquad (5.3\text{-}13)$$

and therefore,

$$N(r) = \frac{N_0 V_0 R_0^2}{r^2 V(r)} \qquad (5.3\text{-}14)$$

The pressure term of (5.3-12) with the help of (5.3-14) becomes,

$$\frac{1}{mN}\frac{d}{dr}(2NkT) = \frac{1}{mN}\frac{d}{dr}\left\{\frac{2N_0V_0R_0^2}{r^2V(r)}kT\right\} = \frac{2k}{m}\left\{\frac{N_0V_0R_0^2}{N(r)}\right\}\frac{d}{dr}\left(\frac{T}{r^2V}\right)$$

$$= \frac{2kr^2V}{m}\frac{d}{dr}\left(\frac{T}{r^2V}\right) = \frac{2kr^2}{m}\frac{d}{dr}\left(\frac{T}{r^2}\right) - \frac{2kT}{mV}\frac{dV}{dr}$$

$$(5.3\text{-}15)$$

Introducing (5.3-15) into (5.3-12) and collecting together the terms with dV/dr we obtain the expression,

$$\left\{V - \frac{2kT}{mV}\right\}\frac{dV}{dr} + \frac{2kr^2}{m}\frac{d}{dr}\left(\frac{T}{r^2}\right) + \frac{GM_0}{r_2} = 0 \qquad (5.3\text{-}16)$$

Equation (5.3-16) is the hydrodynamic equation we must try to solve. We will do it only for the isothermal case, i.e., for $T = T_0 = $ const. Observations show that T is actually a function of r, but as we have seen the thermal conductivity of the solar corona is very high and therefore the temperature varies very slowly with the radial distance. As a result, the assumption of an isothermal corona is a rather good approximation. Performing the differentiation in (5.3-16) using $T = T_0$ we obtain,

$$\left\{V - \frac{2kT_0}{mV}\right\}\frac{dV}{dr} - \frac{4kT_0}{mr} + \frac{GM_0}{r^2} = 0 \qquad (5.3\text{-}17)$$

which can be written also in the form,

$$\left\{V^2 - \frac{2kT_0}{m}\right\}\frac{dV}{V} = \left\{2\left(\frac{2kT_0}{m}\right) - \frac{GM_0}{r}\right\}\frac{dr}{r} \qquad (5.3\text{-}18)$$

We can now introduce the *critical velocity*,

$$V_c^2 = \frac{2kT_0}{m} \qquad (5.3\text{-}19)$$

so that (5.3-18) can be written in the form,

$$(V^2 - V_c^2)\frac{dV}{V} = \left\{2V_c^2 - \frac{GM_0}{r}\right\}\frac{dr}{r} \qquad (5.3\text{-}20)$$

Note that V_c is a velocity very close to the speed of sound V_s because,

$$V_s^2 = \frac{dP}{d\varrho} = \gamma\frac{P}{\varrho} = \gamma\frac{2NkT_0}{mN} = \gamma\frac{2kT_0}{m} = \gamma V_c^2 \qquad (5.3\text{-}21)$$

where γ is the ratio of the specific heats under constant pressure and constant volume which for a monatomic gas is equal to 5/3. Finally, by introducing a *critical radial distance* r_c such that,

$$r_c = \frac{GM_0}{2V_c^2} = \frac{GM_0 m}{4kT_0} \qquad (5.3\text{-}22)$$

we can write the hydrodynamic equation in the following compact form,

$$\left\{ \left(\frac{V}{V_c}\right)^2 - 1 \right\} \frac{dV}{V} = 2\left\{ 1 - \left(\frac{r_c}{r}\right) \right\} \frac{dr}{r} \qquad (5.3\text{-}23)$$

It is of interest to note that r_c can be expressed also in terms of the escape velocity V_e. At a distance R_0, the escape velocity (1.4-10) is given by the relation,

$$V_e = \left(\frac{2GM_0}{R_0}\right)^{1/2} \qquad (5.3\text{-}24)$$

and therefore it follows that,

$$r_c = R_0 \left(\frac{V_e}{2V_c}\right)^2 \qquad (5.3\text{-}25)$$

5.4 The Supersonic Flow of the Solar Wind

In the previous section we have derived the relation (5.3-23) which describes the hydrodynamic equilibrium of the solar corona. For $r < r_c$ and $dr > 0$ the R.H.S. of (5.3-23) is negative. Assuming that in this region we have $V < V_c$, in order to obtain also a negative L.H.S., we must have $dV > 0$, which means that the velocity V increases as we move radially out ($dr > 0$). At $r = r_c$, V must become equal to V_c and for $r > r_c$, i.e., at distances beyond the critical distance, the solar wind must stream outwards with

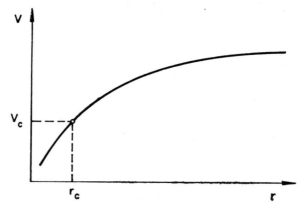

FIGURE 5.4-I   The change of the velocity of the solar wind with radial distance
from the sun

supersonic velocities ($V > V_c$). Figure 5.4-I shows the increase of the velocity of the solar wind with the radial distance from the sun.

As seen from (5.3-22), the critical distance r_c is a function of the gravitational field of the sun, which controls the expansion of the corona up to this critical distance. Beyond this limit, however, the hot plasma of the corona breaks loose of the gravitational force of the sun and flows outwards with supersonic velocity. In this respect, the role of the gravitational force of the sun is similar to the function of the narrow throat of the *de Laval nozzle* (Figure 5.4-II). To understand how the Laval nozzle works, let us consider the hydrodynamic equation (5.3-10) in the absence of a gravity field. Leaving out the gravitational term, we get,

$$\frac{dP}{dr} = -(mNV)\frac{dV}{dr} \tag{5.4-1}$$

The density ϱ of the medium is equal to mN and therefore (5.4-1) can be written in the form,

$$\frac{dP}{\varrho} = -V^2 \frac{dV}{V} \tag{5.4-2}$$

We also have,

$$\frac{dP}{\varrho} = \frac{dP}{d\varrho}\frac{d\varrho}{\varrho} = V_s^2 \frac{d\varrho}{\varrho} \tag{5.4-3}$$

and therefore the equation of hydrodynamic equilibrium in the absence of a gravity field becomes,

$$\frac{d\varrho}{\varrho} = -\left(\frac{V}{V_s}\right)^2 \frac{dV}{V} \tag{5.4-4}$$

The conservation of mass, on the other hand, requires that the total flux of mass through any cross-section A of the nozzle must be constant,

$$\varrho VA = \text{constant} \tag{5.4-5}$$

Taking the logarithm of (5.4-5) and differentiating we get,

$$\frac{d\varrho}{\varrho} + \frac{dV}{V} + \frac{dA}{A} = 0 \tag{5.4-6}$$

which with the help of (5.4-4) yields the relation,

$$\frac{dA}{A} = \left\{ \left(\frac{V}{V_s}\right)^2 - 1 \right\} \frac{dV}{V} \tag{5.4-7}$$

Equation (5.4-7) describes the flow of a fluid in a tube with a variable cross-section A. When the tube narrows down in the direction of the flow (as in the left side of Figure 5.4-II), then dA/A is a negative quantity because dA represents a decrease. For $V < V_s$, i.e. for $\{(V/V_s)^2 - 1\}$ negative, a negative dA/A requires a positive dV/V in (5.4-7). This means that as the cross-section decreases the velocity of the fluid will increase until V will become equal to V_s. At this point we can also make $dA/A = 0$, which means that the tube must stop converging when the speed of the fluid reaches the velocity of sound. If we want V to continue to increase in the supersonic domain, i.e., if we want both dV/V and $\{(V/V_s)^2 - 1\}$ to be positive, we must have also a positive dA/A. This means that after the narrow throat, where V becomes equal to V_s, the cross-section of the tube must start to increase. The de Laval nozzle is often used in rockets (Figure 5.4-III) to propel the combustion gases to supersonic velocities and thus achieve a higher foward thrust for the rocket.

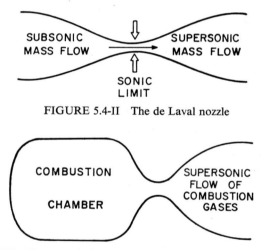

FIGURE 5.4-II   The de Laval nozzle

FIGURE 5.4-III   The use of the de Laval nozzle in rocket engines to achieve supersonic velocities for the ejected combustion gases

By comparing the basic elements that cause the supersonic flow in the solar corona and in the de Laval nozzle, one can readily see that the gravitational force of the sun acts exactly like the converging cross-section of the de Laval nozzle choking the flow up to $r = r_c$. Beyond this point, like beyond the narrow throat of the de Laval nozzle, the fluid breaks free of this constriction and flows away with a steadily increasing supersonic velocity. Let us now return to equation (5.3-23) which describes the hydrodynamic flow in the solar corona,

$$\left\{\frac{V^2}{V_c^2} - 1\right\}\frac{dV}{V} = 2\left\{1 - \frac{r_c}{r}\right\}\frac{dr}{r} \tag{5.4-8}$$

This simple differential equation can be integrated and the supersonic solution can be obtained by requiring that,

$$V = V_c \text{ at } r = r_c \qquad (5.4\text{-}9)$$

Integrating the L.H.S. of (5.4.8) we obtain,

$$\int \left\{ \frac{V^2}{V_c^2} - 1 \right\} \frac{dV}{V} = \frac{1}{V_c^2} \int V dV - \int \frac{dV}{V} = \frac{V^2}{2V_c^2} - \ln V + C_L \qquad (5.4\text{-}10)$$

where C_L is the integration constant. Integrating now the R.H.S. of (5.4-8) we get,

$$2 \int \left\{ 1 - \frac{r_c}{r} \right\} \frac{dr}{r} = 2 \int \frac{dr}{r} - 2r_c \int \frac{dr}{r^2} = 2 \ln r + \frac{2r_c}{r} + C_R \qquad (5.4\text{-}11)$$

where C_R is again the constant of integration. Equating now (5.4-10) and (5.4-11) we obtain,

$$\frac{V^2}{2V_c^2} - \ln V + C_L = \frac{2r_c}{r} + 2 \ln r + C_R \qquad (5.4\text{-}12)$$

Introducing now (5.4-9) into (5.4-11) we obtain the relation,

$$\frac{1}{2} - \ln V_c + C_L = 2 + 2 \ln r_c + C_R \qquad (5.4\text{-}13)$$

which we can subtract from (5.4-12) to eliminate the integration constants. The result is,

$$\frac{V^2}{2V_c^2} - \ln V + \ln V_c - \frac{1}{2} = \frac{2r_c}{r} + 2 \ln r - 2 \ln r_c - 2 \quad (5.4\text{-}14)$$

or,

$$\frac{V^2}{2V_c^2} - \ln \frac{V}{V_c} = \frac{2r_c}{r} + 2 \ln \frac{r}{r_c} - \frac{3}{2} \qquad (5.4\text{-}15)$$

which finally yields the relation,

$$\frac{V^2}{V_c^2} - \ln \frac{V^2}{V_c^2} = 4 \ln \frac{r}{r_c} + \frac{4r_c}{r} - 3 \qquad (5.4\text{-}16)$$

This is the solution of the hydrodynamic equation in the solar corona which is taken to be in isothermal, hydrodynamic equilibrium. For a fully ionized hydrogen plasma at $T_0 = 10^6\,°K$, the critical velocity V_c is,

$$V_c = \left(\frac{2kT_0}{m} \right)^{1/2} = 1.3 \times 10^7 \text{ cm/sec} \qquad (5.4\text{-}17)$$

and therefore the critical distance (5.3-22) is,

$$r_c = \frac{GM_0}{2V_c^2} = 3.9 \times 10^{11} \text{ cm} \simeq 5.6R_0 \tag{5.4-18}$$

This means that the velocity of the solar wind becomes supersonic at a distance of approximately 6 solar radii.

At large distances the term $4r_c/r$ of (5.4-16) tends to zero and therefore it can be neglected. At 1 A.U. $= 215\ R_0$ for example, it is only of the order of 0.1. Furthermore the terms $-\ln (V/V_c)^2$ and -3, are much smaller than the terms $(V/V_c)^2$ and $4 \ln (r/r_c)$ respectively, and in addition they nearly cancel each other. Thus at large radial distances the velocity of the solar wind is given by the expression,

$$V = 2V_c \left(\ln \frac{r}{r_c} \right)^{1/2} \tag{5.4-19}$$

The above equation shows that at large distances the velocity of the solar wind increases very slowly with r because it varies as the square root of a logarithm. This effect is shown in Figure 5.4-I. At the distance of the earth, equation (5.4-19) yields,

$$V_{1\text{A.U.}} = 3.8V_c \simeq 5 \times 10^7 \text{ cm/sec} = 500 \text{ km/sec} \tag{5.4-20}$$

This result is in good agreement with observations which give an average velocity of 350–400 km/sec for the quiet solar wind.

A somewhat better agreement is obtained by including the slow decrease of the temperature with distance and by using the two fluid approach with different thermal conductivities. The improvement, however, with these models is not very spectacular, while the complexity of such computations exceeds by far the simple hydrodynamic problem of a single isothermal fluid we have solved.

In the more sophisticated solutions one must also include the magnetic field of the interplanetary space, which was actually what made the solar wind particles behave like a fluid. If the energy density of the magnetic field $(H^2/8\pi)$ is much larger than the thermal energy density of the fully ionized plasma $(2NkT)$, then the critical velocity V_c, which is closely connected to the speed of sound V_s (5.3-21), must be replaced by the *Alfvén velocity* V_A. The magnetic field lines that are frozen into the plasma behave like stretched elastic strings and any perturbation of the magnetic field induced by a motion of the plasma travels rapidly along the field lines in the form of waves which are called *Alfvén* waves. In the case of acoustic waves, the kinetic energy per unit volume of the waves propagating with velocity V_c is closely related to the thermal energy per unit volume,

$$\frac{1}{2}\varrho V_c^2 = NkT \qquad (5.4\text{-}21)$$

and this relation gives the expression for V_c which we used before (5.3-19),

$$V_c = \left(\frac{2NkT}{\varrho}\right)^{1/2} = \left(\frac{2NkT}{mN}\right)^{1/2} = \left(\frac{2kT}{m}\right)^{1/2} \qquad (5.4\text{-}22)$$

Similarly, in the case of the hydromagnetic waves the kinetic energy of the Alfvén waves per unit volume is equal to the energy density of the magnetic field,

$$\frac{1}{2}\varrho V_A^2 = \frac{H^2}{8\pi} \qquad (5.4\text{-}23)$$

and this relation gives the expression for the Alfvén velocity,

$$V_A = \frac{H}{(4\pi\varrho)^{1/2}} = \frac{H}{(4\pi Nm)^{1/2}} = c\,\frac{\left(\dfrac{eH}{2\pi mc}\right)}{\left(\dfrac{e^2 N}{\pi m}\right)^{1/2}} = c\left(\frac{f_{Hi}}{f_{Ni}}\right) \qquad (5.4\text{-}24)$$

where f_{Ni} is the ion plasma frequency and f_{Hi} is the ion cyclotron frequency.

When the energy density of the magnetic field is comparable to the thermal energy density then the waves are neither pure acoustic nor pure Alfvénic, but rather a combination of the two. These are called magneto-acoustic waves and there are actually many varieties of them. In one of the less complicated cases the velocity of the magneto-acoustic waves is given by the expression,

$$V_{MA} = (V_A^2 + V_s^2)^{1/2} \qquad (5.4\text{-}25)$$

At a distance of approximately 1 A.U., where $T \sim 10^5\,°K$, $N \simeq 5$ protons/cm^3 and $H \simeq 6 \times 10^{-5}$ Gauss, the velocity of sound is,

$$V_s = \left(\gamma\,\frac{2kT}{m}\right)^{1/2} \simeq 5 \times 10^6\ \text{cm/sec} = 50\ \text{km/sec} \qquad (5.4\text{-}26)$$

and the Alfvén velocity is,

$$V_A = \frac{H}{(4\pi Nm)^{1/2}} \simeq 6 \times 10^6\ \text{cm/sec} = \text{km/sec} \qquad (5.4\text{-}27)$$

which shows that they actually are of the same order of magnitude. The velocity of the solar wind, on the other hand, which is of the order of 400 km/sec, is much higher than either of the two and therefore we are dealing with a supersonic flow with a Mach number of the order of 10. As a result a shock wave develops when the solar wind encounters the obstacle of the magnetic cavity of the earth. This shock wave in front of the magnetosphere is no different than the shock wave produced by a body travelling at supersonic velocities through a stationary medium.

5.5   The Interplanetary Magnetic Field

From Maxwell's equations follows that a highly conducting plasma, as the hot plasma of the solar corona, will drag along any embedded magnetic field. Thus the loops of the solar magnetic field, which arch into the corona and are anchored in the active regions of the low latitudes, will be stretched out, near the ecliptic, in the interplanetary space by the outflowing solar wind. While the loops of the solar magnetic field are pulled out radially by the solar wind, the sun rotates about its axis winding up the stretched out field lines. As a result the interplanetary magnetic field near the ecliptic, which is anchored in the sunspot zone, assumes a spiral configuration (Figure 5.5-I). If τ is the sidereal (relative to the stars) period of rotation of the sun, in a time interval dt, the sun will rotate by an angle $d\omega$,

$$d\omega = \frac{2\pi}{\tau}\, dt \tag{5.5-1}$$

In the same time interval the solar wind particles move in the radial direction a distance dr,

$$dr = V\, dt \tag{5.5-2}$$

where V is the velocity of the solar wind. As seen from Figure 5.5-I,

$$r\, d\omega = dr\, \tan \psi \tag{5.5-3}$$

and therefore from the above equations it follows that,

$$\tan \psi = \frac{r\, d\omega}{dr} = \frac{r\, d\omega}{V\, dt} = \frac{2\pi r}{\tau V} \tag{5.5-4}$$

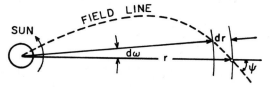

FIGURE 5.5-I   The geometry of the interplanetary magnetic field is an Archimedean spiral

The angle ψ is called the *streaming angle*. Taking $\tau \simeq 25$ days $\simeq 2.2 \times 10^6$ sec, we find that for typical velocities of the solar wind at the distance of the earth,

$$\tan \psi_0 \simeq 1 \tag{5.5-5}$$

i.e.,

$$\psi_0 \simeq 45° \tag{5.5-6}$$

From (5.5-4) we also have that,

$$d\omega = \left(\frac{2\pi}{\tau V}\right) dr \tag{5.5-7}$$

which can be integrated to give,

$$\omega(r) = \frac{2\pi}{\tau V}(r - r_0) + \omega_0 \tag{5.5-8}$$

where ω_0 is the angle ω at $r = r_0$. Equation (5.5-8) is the equation of the *Archimedean spiral* which describes the geometry of the magnetic field lines in the interplanetary space. This is also called the *garden hose effect* because the lines of the magnetic field, like the water jet of a rotating garden hose, form a curved spiral while the solar wind particles, like the water droplets of the water jet, move always in the radial direction. A diagrammatic representation of the phenomenon is shown in Figure 5.5-II.

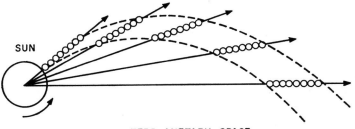

INTERPLANETARY SPACE

FIGURE 5.5-II   The "gardenhose" effect of the solar wind and the inter-planetary magnetic field

It should be noted that though the solar wind particles move away from the sun in the radial direction, a terrestrial observer has the impression that they are coming from a direction approximately 5° to the west of the sun. This is due to an *aberration effect* which is produced by the motion of the earth at right angles to the sun-earth direction. The phenomenon is the same as in the case of a moving observer who thinks that the rain is falling at a slant angle ϕ, while a stationary observer sees the rain falling vertically

(Figure 5.5-III). In the case of the solar wind the tangential velocity with which the earth moves around the sun is $V_0 \simeq 30$ km/sec, while the radial velocity of the solar wind is $V \simeq 400$ km/sec. Hence,

$$\tan \phi = \frac{V_0}{V} = \frac{30}{400} = 0.075 \qquad (5.5-9)$$

and therefore,

$$\phi \simeq 4.3° \qquad (5.5-10)$$

FIGURE 5.5-III   Due to the aberration effect, a moving observer thinks that the rain is falling in a slanted direction while a standing observer sees it falling vertically

If we know the geometry of the magnetic field in the interplanetary space, we can compute its intensity using the conservation of the total magnetic flux in a tube of force. This principle requires that the product of the area of any given cross-section of the tube times the intensity of the component of the magnetic field normal to this area be always constant,

$$A_0 H_0 = A_r H_r = \text{constant} \qquad (5.5-11)$$

As we have seen, the interplanetary magnetic field makes an angle ψ with the radial vector and therefore if A_r is a cross-section of the tube normal to the radial vector, then H_r must be the radial component of the magnetic field,

$$H_r = H \cos \psi = \frac{H}{(1 + \tan^2 \psi)^{1/2}} \qquad (5.5-12)$$

If H_0, on the other hand, is the radial magnetic field near the surface of the sun, then as seen from Figure 5.5-IV, the corresponding cross-section A_0 of the tube is related to A_r through the expression,

$$\frac{A_0}{A_r} = \left(\frac{R_0}{r}\right)^2 \qquad (5.5-13)$$

where R_0 is the radius of the sun. From the above equations we get,

$$H = H_r(1 + \tan^2 \psi)^{1/2} = H_0 \left(\frac{A_0}{A_r}\right)(1 + \tan^2 \psi)^{1/2}$$

$$= H_0 \left(\frac{R_0}{r}\right)^2 \left\{1 + \left(\frac{2\pi r}{\tau V}\right)^2\right\}^{1/2} \qquad (5.5\text{-}14)$$

At the distance of the earth, where $\tan \psi \simeq 1$ and $r/R_0 \simeq 215$, taking $H_0 \simeq 2$ Gauss we find,

$$H_{1\text{A.U.}} \simeq 6 \times 10^{-5} \text{ Gauss} = 6\gamma \qquad (5.5\text{-}15)$$

which is in agreement with actual observations.

FIGURE 5.5-IV   The conservation of the total magnetic flux in a tube of force of the interplanetary magnetic field

Studies of the interplanetary magnetic field between the orbits of the earth and Mars with Mariner (Colemen et al., 1969), have shown that the intensity of the total magnetic field varies with r in agreement with (5.5-14). The same observations, however, have found some discrepancies with the tangential and the radial component of the interplanetary field. In general the magnitude of the interplanetary magnetic field tends to be much more constant and predictable than its direction. There is also a strong positive correlation between the intensity of the solar wind, as expressed by the Kp index of the geomagnetic activity, and the strength and variability of the interplanetary magnetic field.

During periods of low solar activity, the interplanetary space is divided into 4 sectors (Figure 5.5-V) where the lines of force reverse direction. These sectors co-rotate with the sun and have been observed to persist for several solar rotations before they are replaced by a different but similar pattern. It is interesting to note that the solar magnetic field in the photosphere, through its inward or outward direction, divides the surface of the sun in the same number of sectors. These sectors cross the central meridian of the solar disc 4 to 5 days before the corresponding sector of the interplanetary magnetic field is detected at the distance of the earth. This confirms the solar origin of the interplanetary magnetic field and the fact that

10*

it is actually pulled away from the sun by the solar wind which with a typical velocity of 350–400 km/sec covers the sun-earth distance in about 4 to 5 days.

The different sectors are separated by very thin neutral sheets (less than 1% of the width of the sector) where the magnetic field undergoes a 180° reversal. The magnetic field in each sector reaches a maximum approximately after 0.3 of the width of the sector following the preceding neutral sheet. Of particular interest is the correlation of the interplanetary sector structure with persistent streams of solar cosmic rays which show a recurrence with a period of about 27 days. A similar correlation is observed with recurrent variations of the solar wind as detected by the resulting magnetic activity on the earth. During periods of high solar activity the interplanetary medium is divided into many more sectors, which as a result are less well defined.

The plasma density of the interplanetary space shows usually a minimum near the neutral sheet but increases rapidly behind the preceding boundary of the sector. As a second order effect, the interplanetary medium shows a rich, fine scale filamentary structure in which individual flux tubes appear to be directly connected to the sun. Small scale irregularities in the plasma density of the interplanetary medium have also been detected (Hewish, 1958) from the rapid fluctuations observed in the radio emission of different strong radio sources, such as the Crab Nebula in Taurus. These fluctuations are called *interplanetary scintillations*. The period of these scintillations is

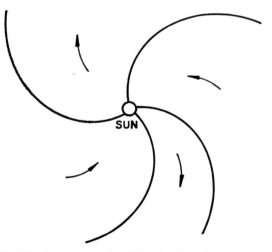

FIGURE 5.5-V   The four sectors in which the interplanetary space is divided during quiet solar conditions. The interplanetary magnetic field reverses direction at the boundaries of the sectors

typically of the order of 1 second which means that the typical size of the
solar wind irregularities is of the order of 400 km ($L = Vt$). The fast period
of the interplanetary scintillations sets them apart from the ionospheric
scintillations which typically have periods of the order of 1 minute.

The solar wind at the distance of the earth is characterized by the following typical values:

Velocity	350–400 km/sec,
Direction	Very nearly radial,
Plasma density	5–10 protons/cm³ (plus equal number of electrons),
Alpha particles	∼5% of the number of protons, but often changing,
Ion flux	2×10^8 ions cm^{-2} sec^{-1},
Temperature of electron gas (T_e)	1.5×10^5°K,
Temperature of proton gas (T_p)	5×10^4°K,
Temperature of α-particles (T_α)	1.5–2×10^5°K,

During periods of high solar activity the velocity of the solar wind can
increase up to 900 km/sec, the direction can vary by as much as $\pm 15°$
from the radial, and the ion density can reach up to 80 protons/cm³. The
fraction of helium nuclei can occasionally become as high as 15%, while
the solar wind flux can reach values as high as 10^{10} ions cm^{-2} sec^{-1}. During
highly active periods the temperatures T_e and T_p tend to equalize and can
both become as high as 9×10^5°K. Actually there is a direct correlation
between the temperature of the proton gas T_p and the velocity of the solar
wind V which can be described by the semi-empirical relation,

$$T_p^{1/2} = (0.036 \pm 0.003) V - (5.54 \pm 1.50) \qquad (5.5\text{-}16)$$

where T_p is expressed in kilo-degrees Kelvin and V in km/sec.

The value of V, which is directly related to the activity on the sun, can
also be inferred from the prevailing geomagnetic activity through the
relation,
$$V = (8.44 \pm 0.74) \Sigma Kp + (330 \pm 17) \text{ km/sec} \qquad (5.5\text{-}17)$$

where ΣKp is the daily sum of the 3-hour global index of geomagnetic
activity. It should be mentioned that the interplanetary magnetic field
near the earth can also show a large increase up to 40γ following a peak
of solar activity. During such an increase the cosmic ray flux on earth shows
a sharp decrease because the enhanced interplanetary magnetic field provides an additional shielding for the earth against the cosmic rays. This
sudden change in the cosmic ray flux is called a *Forbush decrease*, after
Scott E. Forbush (1956) who was the first to observe this effect.

It is not very clear how far the solar wind extends. It is believed, however, that at a distance of approximately 50–100 A.U. the pressure of the solar wind ($2\mathrm{Nm}V^2$) will be balanced by the pressure of the interstellar magnetic field ($H^2/8\pi$) which has an intensity of about 1γ. At this distance the supersonic solar wind will undergo a shock transition to subsonic flow forming the boundary of the domain of the solar system. This boundary could be called the *coronapause*, since the solar wind represents the natural extent of the solar corona.

It is interesting to add that the solar system moves toward a point between the constellations of Lyra and Hercules with a velocity of approximately 20 km/sec relatively to its neighboring stars and therefore also relatively to the surrounding interstellar medium. As a result it is quite possible that the coronapause of the sun, like the magnetopause of the earth, is compressed in the forward direction and stretched out to a tail in the backward direction.

The total mass ΔM_0 that the sun has lost over its entire history through the solar wind is approximately equal to,

$$\Delta M_0 \simeq SAtm \simeq 2 \times 10^{29}\mathrm{gr} \qquad (5.5\text{-}18)$$

where $S \simeq 3 \times 10^8$ protons cm^{-2} sec^{-1} is an average flux for the solar wind at the distance of the earth, $A \simeq 2.7 \times 10^{27}$ cm^2 is the area of a sphere 1 A.U. in radius, $t \simeq 4.7$ billion years $\simeq 1.5 \times 10^{17}$ sec is the age of the sun, and $m = 1.6 \times 10^{-24}$ gr is the mass of a proton. As seen from (5.5-18), ΔM_0 is only 0.0001 of a solar mass ($M_0 = 2 \times 10^{33}$ gr) and therefore is a negligible quantity for the sun as a whole. It constitutes, however, an important renewing process for the solar atmosphere.

5.6   Interplanetary Dust

In addition to ionized particles (solar wind) and magnetic fields, the interplanetary space contains also a considerable amount of solid matter. Most of this matter is in the form of small particles which are called the *interplanetary dust*. The entire assortment of all the solid pieces in the interplanetary space, including some that are tens of meters in size, is called the *meteoritic complex*.

The oldest source of information about the interplanetary dust is the *zodiacal light*. This is a band of dim, diffuse light which occasionally can be seen along the constellations of the zodiac, or more accurately along the plane of the ecliptic. When seeing conditions are favorable, the zodiacal light can be seen after sunset or before sunrise forming a luminous, cone-shaped patch above the horizon, in the area where the sun sets or rises.

There is also a small intensification of the zodiacal light in the antisolar direction which is called the *gegenschein*.

Casini had suggested nearly 300 years ago that the zodiacal light is caused by the reflection of the sunlight from dust particles in the interplanetary space. This is also the presently accepted explanation. A few decades ago it was thought that scattering from interplanetary electrons was also a contributing factor because it was found that nearly 20% of the zodiacal light is polarized. Computations had shown that an electron density of 500–1000 el/cm^3 could account for the polarization of the zodiacal light. Now, however, we know that the electron density of the interplanetary space is only about 10 el/cm^3 and therefore the idea that a substantial part of the zodiacal light is due to electron scattering has been abandoned.

In Section 4.2 we discussed the F-corona. Observations show that the light intensity of the F-corona joins smoothly with the zodiacal light and therefore it is tempting to think that the F-corona is also the result of scattering from dust particles in the vicinity of the sun. The F-corona, however, reaches a maximum at a distance of only a few solar radii from the sun, where there are no dust particles. The reason is that at distances less than 10–20 solar radii all dust particles are rapidly evaporated by the heat of the sun. This problem was finally resolved in the 1940's by Allen and by Van de Hulst who concluded that the diffuse light which we think comes from near the sun is actually produced away from the sun in the space between the sun and the earth by diffraction, rather than by scattering, from dust particles.

Light rays change very little their paths through diffraction and therefore this process works only at small angles to the sun, which we falsely had interpreted as small distances from the sun. Thus the F-corona is only an optical effect and not something that surrounds the sun. At larger angles to the sun, the diffraction effect decreases rapidly and is progressively replaced by reflection (scattering) from dust particles. By now, however, the angles are such that the particles do not have to be any closer to the sun than 10 or 20 solar radii. The gradual change from diffraction to scattering accounts for the decrease in the intensity and the smooth transition from the F-corona to the zodiacal light.

The existence of dust particles in the interplanetary space is also strongly suggested by the meteors that are continuously striking the earth. A *meteor* is the luminous effect produced by the entry of a particle from space into the earth's atmosphere. Typical entry velocities are of the order of 30 km/sec. The term *meteoroid* describes these particles while they are still in the interplanetary space. The inner cores of meteoroids heavier than about one kilogram can survive the passage through the earth's atmosphere and reach the ground. Meteoroids found on the surface of the earth are called *meteorites*. The largest meteorites found on earth (Hoba West, Ahnighito, etc.) weigh about 50 tons but there is evidence for some even larger ones. The number of meteorites N with mass larger than m in kilograms that fall over the entire earth per day is approximately,

$$N = \frac{260}{m} \qquad (5.6\text{-}1)$$

Of course a meteorite of mass m was originally a meteoroid with a mass of about $5m$. This is called the *ablation correction* and accounts for the mass that was evaporated from the red hot meteoroid while crossing the earth's atmosphere. Meteorites are classified essentially into irons (mostly metal Fe and Ni) with a density of 8 gr/cm³, and stones (mostly silicates) with a density of 3.5 gr/cm³. Only one in every ten meteorites is an iron meteorite but because they are recognized much easier, they represent the largest portion of the meteorites found, which is less than 1 % of all the meteorites falling on earth.

Meteoroids smaller than about 10 microns can also avoid complete evaporation in the earth's atmosphere because their small diameter (L) gives them a large surface (A) to mass (M) ratio ($A/M = L^2/L^3 = 1/L$), which allows them to radiate away most of the frictional heat which is generated during their deceleration in the terrestrial atmosphere. These tiny particles are finally stopped at an altitude of about 100 km, and from there they make a free fall to the surface of the earth with a limiting velocity given by Stoke's law. When they reach the surface of the earth, these minute particles are called *micrometeorites*. It is very difficult to compute now the influx of micrometeorites, because our modern industries add to the air a much larger number of small particles which ultimately fall also to the ground and confuse any counting attempts. Scientists, however, are trying to study the micrometeorite rates of the past by examining deep core samples from the ocean floors and deep snow samples from the arctic regions. It is estimated that the accretion rate of micrometeoritic material by the earth might be as high as 100 tons per day.

Meteors have been studied mostly by the light trail which they leave on the night sky. Meteors also ionize the atmosphere along their trajectory which as a result can reflect radar signals. These are called *radar meteors* and can be studied even during the day. Diurnal studies of meteor events show that the occurrence of meteors reaches a maximum near 6:00 a.m. local time. As seen from Figure 5.6-I, sunrise occurs in the regions that are in the front part of our planet as the earth moves along its orbit around the sun. It is not surprising, therefore, that the front part of the earth will sweep more interplanetary dust particles than any other area of the planet.

The larger meteoroids, which produce most of the meteorites we recover, are believed to be of asteroidal origin, i.e., the results of the grinding of the asteroids through collisions. The smaller meteoroids, on the other hand, which produce meteors but not meteorites and often have a fluffy consistency with a density of only about $0.4 \, gr/cm^3$, are believed to be of cometary origin. These particles, in other words, came from the ice and dust nuclei of comets which disintegrated either partially or totally. This belief is supported also by intense meteor showers (Leonids, Perseids, etc.) which occur when the earth crosses through the dust-filled orbits of past comets. On an average night, one can see about 10 meteors per hour, but during a very intense meteor shower the hourly rate can increase to several thousands.

The interplanetary particles besides being of asteroidal or cometary origin, they could also be primordial debris left over from the formation of the solar system. Another possible source is interstellar dust accreted by the solar system during its travellings through the galaxy. Neither of these two sources is considered now a very important factor, though in the past they received at times considerable support. One of the reasons why small dust particles could not have remained in their present orbits since the formation of the solar system is the very interesting *Poynting–Robertson effect* which is produced by the action of the solar radiation on the dust particles. This effect was discovered by Poynting in 1903 and was formulated on a rigorous relativistic basis by Robertson in 1937.

The sunlight exerts a pressure on the interplanetary dust particles which cancels part of the gravitational attraction of the sun. Still, if the gravitational force is stronger than the radiation force, the particles should remain in well-determined orbits around the sun. Due to the aberration effect, however, an orbiting particle receives the solar photons at an angle to the radial vector. As a result the radiation force is not exactly a radial force and there-

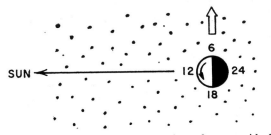

FIGURE 5.6-I   The sweeping of a large number of meteoroids by the front part of the earth as the earth moves in its orbit around the sun

fore it can be analyzed into a radial and a tangential component. The tangential component opposes the motion of the particle which as a result loses slowly angular momentum and spirals inwards toward the sun. A spherical particle of radius S and density ϱ, which has an orbit of a perihelion distance q (in A.U.) and eccentricity e, will spiral into the sun in a time interval t given by the expression,

$$t = C_e \varrho S^2 \times 10^7 \text{ years} \qquad (5.6\text{-}2)$$

where C_e is equal to 0.7 for $e = 0$ (circular orbit), 1.9 for $e = 0.5$, and 7.3 for $e = 0.9$. Consequently all the small particles at distances of a few A.U. from the sun will spiral and vanish into the sun in time intervals that are much shorter than the age of the solar system ($\sim 4.7 \times 10^9$ years). The micron rangle particles that are now responsible for the zodiacal light could have been present 4.7×10^9 years ago only if they were approximately 500 A.U. away from the sun.

The Poynting–Robertson effect will not bring all the particles toward the sun because for some particles the radiation force F_R is actually greater than the gravitational force F_G and the solar radiation will push these particles out of the solar system. The computation, however, is not as simple as it is presented in some elementary texts. The radiation pressure is equal to the radiation density U, which is equal to the flux of the black body radiation divided by the velocity of light. Hence the total force F_R due to radiation pressure on a spherical particle of radius S, a distance r from the sun is,

$$F_R = \pi S^2 U(r) = \pi S^2 \frac{1}{c} \sigma T_0^4 \left(\frac{R_0}{r}\right)^2 \qquad (5.6\text{-}3)$$

On the other hand, if ϱ is the density of the particle, the gravitational force F_G from the sun acting on it is,

$$F_G = \frac{GM_0 m}{r^2} = \frac{GM_0}{r^2} \left(\frac{4}{3} \pi S^3 \varrho\right) \qquad (5.6\text{-}4)$$

If $F_G > F_R$ the particle will remain in orbit in the solar system and the Poynting–Robertson effect will slowly pull it closer to the sun. If, however, $F_R > F_G$ then the particle might be pushed out of the solar system. The ratio β of the two forces is,

$$\beta = \frac{F_R}{F_G} = \frac{3\sigma T_0^4 R_0^2}{4GM_0 c} \frac{1}{S\varrho} = \frac{5.7 \times 10^{-5}}{S\varrho} \qquad (5.6\text{-}5)$$

where S is in cm and ϱ in gr/cm^3. A careful examination of equation (5.6-5) shows that for $\varrho \simeq 1$, F_R becomes larger than F_G when S is smaller than 5.7×10^{-5} cm = 0.57 microns = 5,700 Å. But when the size of the particle is approximately equal to wavelength λ of the radiation, the classical theory breaks down and must be replaced by a more sophisticated treatment of the light scattering problem. Such a theory was first developed by Mie in 1908 and showed that β depends also on the ratio S/λ and the dielectric properties of the particles. For this reason, a β in (5.6-5) greater than unity does not necessarily mean that the force due to the radiation pressure is larger than the gravitational force. As a matter of fact for some materials this never happens even for very small values of S. Other materials, however, such as spherical iron particles with radii $S < 0.5\ \mu$, are ultimately pushed out of the solar system.

Meteoroids moving with high velocities in the interplanetary space present a potential danger for manned space flights. The magnitude of the danger depends on the number density, the velocity distribution, the orbits, the size distribution, the consistency, and the structural strength of these particles. In order to provide adequate shielding for all manned space flights, without at the same time adding unnecessary weight to the space crafts, it is essential to have quantitative information on all these parameters. Meteor studies and zodiacal light observations can provide only a limited amount of data along these lines. For this reason many rockets, satellites and deep space probes have been launched to collect direct information about the meteoritic complex in the interplanetary space around the earth. The devices used in these flights belong to one of the following two general groups.

One method uses a microphone detection system for acoustic impact counting. This system consists basically of a metallic sensor plate with a tranducer attached to it. The electrical signals produced by the piezoelectric crystals are essentially proportional to the momentum of the impact particles. Such sensors were carried by the Mariner spacecrafts that flew to Mars and Venus and measured the dust particles in the region between 0.7 and 1.5 A.U. In the same general group one must include also the photometric impact sensors which count light flashes caused by particle impacts on a sensitized surface.

The other basic technique used, is counting the rate at which interplanetary particles penetrate metal sheets of various thicknesses. Each penetration is recorded through the shorting of a capacitor or some other equivalent method. Devices of this type were carried in space by several of the Explorer satellites (e.g. Explorer XXIII) and the three Pegasus satellites. In the same general group we must include also cratering counting devices which study the craters produced on special plates exposed to the space environment.

The proper interpretation of some of these data remains still uncertain. In addition, these direct counting techniques are often confused by large numbers of particles which fly out of the satellite. As a result we still cannot claim that we have a complete and accurate picture of all the important parameters of the meteoritic complex of the interplanetary space. Some of the experiments found a much higher concentration of dust particles near the earth, which suggested that the earth was engulfed by a dust cloud. This could have been either an effect of the earth's gravitational field, or that these particles had been ejected from the moon by the splashing of large meteorites or through volcanic eruptions. Other experiments, however, seem to find little difference between measurements near and far away from the earth, and therefore the subject remains still unclear.

Figure 5.6-II summarizes our present knowledge, from all the available sources, of the mass distribution of micrometeorites near the orbit of the earth (Kerridge, 1970). The average density of the interplanetary dust at a heliocentric distance of about 1 A.U. is roughly 10^{-22} gr/cm^3. Most of this mass is concentrated in particles with a radius around $100\,\mu$ and a mass of the order of 10^{-5} gr. A substantial portion of the total mass might also be in the micron and the sub-micron component. Our present values for the total mass and its distribution, the densities, the velocity distribution and the spacial distribution of the dust particles with distance from the sun are still uncertain by one or more orders of magnitude (Bandermann and Singer, 1969). More sophisticated experiments will undoubtably improve the accuracy of these measurements in the coming years and will provide a more complete description of the different parameters which characterize the meteoritic complex of the interplanetary space.

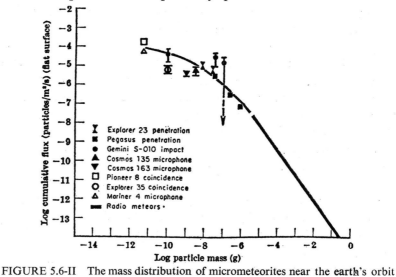

FIGURE 5.6-II   The mass distribution of micrometeorites near the earth's orbit
(Kerridge, 1970)

5.7 Bibliography

A. Books for Further Studies

1. *Interplanetary Dynamical Processes*, E. N. Parker, John Wiley & Sons, New York, N. Y., 1963.
2. *Solar System Astrophysics*, J. C. Brandt and P. W. Hodge, McGraw Hill, New York, N. Y., 1964.
3. *Plasma Dynamics*, E. H. Holt and R. E. Haskell, The Macmillan Co., New York, N. Y., 1965.
4. *Physics of Fully Ionized Gases*, L. Spitzer Jr., Interscience, New York, N. Y., 1962.
5. *Meteors, Comets and Meteorites*, G. S. Hawkins, McGraw Hill, New York, N. Y., 1964.
6. *The Solar Wind*, ed. by R. J. Mackin Jr., and M. Neugebauer, Pergamon Press, Long Island City, N. Y., 1966.
7. *Introduction to Space Science*, ed. by W. N. Hess and G. D. Mead, Gordon and Breach, New York, N. Y., 1968.
8. *Research in Geophysics*, Vol. I, ed. by H. Odishaw, the MIT Press, Cambridge, Mass., 1964.
9. *Space Physics*, ed. by D. P. LeGalley and A. Rosen, John Wiley & Sons, New York, N. Y., 1964.
10. *Satellite Environment Handbook*, 2nd Edition, ed. by F. S. Johnson, Stanford University Press, Stanford, Calif., 1965.
11. *Introduction to Solar Terrestrial Relations*, ed. by J. Ortner and H. Maseland, Gordon and Breach, New York, N. Y., 1965.
12. *Solar Terrestrial Physics*, ed. by J. W. King and W. S. Newman, Academic Press, New York, N. Y., 1967.
13. *The Zodiacal Light and the Interplanetary Medium*, ed. by J. L. Weinberg, NASA SP-150, U. S. Gov. Print. Office, Washington, D. C., 1967.
14. *Introduction to the Solar Wind*, J. C. Brandt, W. H. Freeman and Co., San Francisco, Calif., 1970.
15. *Space Physics*, R. S. White, Gordon and Breach, New York, N. Y., 1970.

B. Articles in Scientific Journals

Bandermann, L. W. and S. F. Singer, Interplanetary dust measurements near the earth, *Rev. of Geophys.*, **7**, 759, 1969.

Colleman, P. J. *et al.*, The radial dependence of the interplanetary magnetic field: 1.0 to 1.5 AU, *J. Geophys. Res.*, **74**, 2826, 1969.

Dessler, A. J., Solar wind and interplanetary magnetic field, *Rev. of Geophys.*, **5**, 1, 1967.

Forbush, S. E., World wide cosmic ray variations, 1937–1952, *J. Geophys. Res.*, **59**, 525, 1954.

Hundhausen, A. J., Composition and dynamics of the solar wind plasma, *Rev. Geophys. and Space Phys.*, **8**, 729, 1970.

Hewish, A., The scattering of radio waves in the solar corona. *Mon. Not. Roy. Astron. Soc.*, **118**, 534, 1958.

Kerridge, J. F., Micrometeorite environment at the earth's orbit, *Nature*, **228**, 616, 1970.

Parker, E. N., Dynamical theory of the solar wind, *Space Sci. Rev.*, **4**, 666, 1965.

CHAPTER 6

SOLAR-TERRESTRIAL RELATIONS

6.1 Introduction

In the first three chapters we have studied the atmosphere, the ionosphere, and the magnetosphere that surround our planet. The conditions prevailing in these regions depend directly on the level of the solar activity and therefore in Chapter 4 we studied the active sun. As we have seen, the solar emission can be analyzed into three components (Figure 6.1-I). The first one is the steady component representing the constant flux of solar radiation which determines the basic features (temperature, electron density, size of the magnetic cavity, etc.) of the terrestrial domain. It also determines the basic nature (solar wind, magnetic fields, etc.) of the interplanetary space, which we have studied in Chapter 5.

The second component of the solar emission is the slowly varying component, which is superimposed on the steady level of the solar emission and is responsible for the long-term periodic changes in the physical properties of the terrestrial atmosphere, ionosphere, and magnetosphere. The period of these variations is the 11 year cycle of the solar activity. One could include also in this category a 27 day periodicity due to the rotation

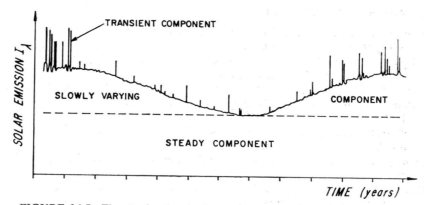

FIGURE 6.1-I The steady, the slowly varying, and the transient components of solar emission, at a certain wavelength λ. The relative intensities of these components are wavelength dependent

of the sun. The periodic reappearance of active centers on the face of the sun often causes a 27 day fluctuation in the properties of the terrestrial domain and the conditions of the interplanetary space in the vicinity of the earth. Of course the atmosphere and the ionosphere at a given latitude undergo also some seasonal variations, but these are due to the rotation of the earth around the sun and therefore they are not related to the level of solar activity.

The third component of the solar emission is a transient, explosive component which lasts only a few minutes or at most a few hours. Our ability to forecast the occurrence and the magnitude of these events is still very limited, and because some of them can cause serious problems such as radio blackouts, radiation hazards for men in space, etc., a well coordinated effort is made now to improve our short term and long term forecasting capabilities for solar flare events.

The effects of the solar radiation, and especially of its transient component, on the atmosphere, the ionosphere, and the magnetosphere of the earth are studied by a special branch of space physics. This new branch is called *solar-terrestrial relations* and is the offspring of a successful pairing of geophysics with solar physics. Because of their many scientific and practical applications, solar-terrestrial relations have been the subject of voluminous research during the past decades. Rockets and satellites have given a tremendous impetus to these studies during the last two decades (see Appendix II) because they made possible the in situ observation of the different disturbances in the atmosphere, ionosphere, and magnetosphere of the earth. In addition they have provided the means to intercept and study directly in the interplanetary space the energetic particles, the plasma clouds, and the x-ray bursts that are responsible for these disturbances.

The study of a particular solar-terrestrial relation starts usually with the discovery of a transient geophysical phenomenon which shows a good correlation with solar activity. The new effect is often given a long descriptive name, e.g., a *sudden ionospheric disturbance*, which is later identified by its initials written in capital letters, e.g., SID in the above-mentioned case. The general disturbance of the terrestrial domain appears as a different effect when the latitude or the altitude is changed. Furthermore a disturbance can be detected as a different effect through the many modes of observation (radio, magnetic, optical, etc.). As a result the field of solar-terrestrial physics abounds in such effect and has a great plethora of acronyms to describe them. We will study the most important of these effects in the different sections of this chapter.

Before proceeding with the study of the different solar-terrestrial relation‘s it is useful to discuss the quantities of energies involved in the overall event. The total amount of solar radiation falling on the earth per second is,

$$W_R = S_0 A_E = 1.8 \times 10^{24} \text{ erg/sec} \qquad (6.1\text{-}1)$$

where S_0 is the solar constant $(1.4 \times 10^6 \text{ erg cm}^{-2} \text{ sec}^{-1})$ and A_E is the cross-section of the earth $(1.3 \times 10^{18} \text{ cm}^2)$. This quantity changes by less than 1% during peaks of solar activity, though of course the intensity in certain regions of the spectrum, such as in the x-ray region around 1 Å, might increase by several orders of magnitude.

In addition to the electromagnetic radiation, the earth with its magnetosphere intercepts also the corpuscular radiation from the sun (solar wind). The energy flux of the solar wind during quiet periods is,

$$F_s = NV\left(\frac{1}{2} m_p V^2\right) \simeq 0.2 \text{ erg cm}^{-2} \text{ sec}^{-1} \qquad (6.1\text{-}2)$$

where $N \simeq 5 \text{ protons/cm}^3$ and $V = 350\text{–}400 \text{ km/sec}$. The cross-section of the magnetosphere A_M, with a radius of $20\text{–}25$ earth radii, is approximately $7 \times 10^{20} \text{ cm}^2$, and therefore during quiet solar conditions the total amount of solar corpuscular energy falling per second on the terrestrial magnetosphere is,

$$W_s = F_s A_M \simeq 1.5 \times 10^{20} \text{ erg/sec} \qquad (6.1\text{-}3)$$

During peaks of solar activity, unlike the W_R of (6.1-1), W_s can increase by one or even two orders of magnitude, producing strong transient effects on earth. Typical values of the velocity and the particle density of the solar wind following a moderately strong solar flare are: $V = 750 \text{ km/sec}$ and $N = 25 \text{ protons/cm}^3$ which yield,

$$F_s \simeq 8 \text{ erg cm}^{-2} \text{ sec}^{-1} \qquad (6.1\text{-}4)$$

This is approximately 40 times the value for quiet conditions (6.1-2), and raises the total energy intercepted per second to a value,

$$W_s \simeq 6 \times 10^{21} \text{ erg/sec} \qquad (6.1\text{-}5)$$

The typical duration of such an event is $\sim 10^3$ sec and therefore the total energy available is,

$$E_s \simeq 6 \times 10^{24} \text{ ergs} \qquad (6.1\text{-}6)$$

Of this amount only about $10^{21}\text{–}10^{22}$ ergs are dissipated on the earth. This energy goes mostly to the following three activities at the rates shown below,

a. Auroras $(\sim 10^{18} \text{ erg/sec})$
b. Heating of the ionosphere $(\sim 10^{18} \text{ erg/sec})$
c. Magnetic disturbances $(10^{18}\text{–}10^{19} \text{ erg/sec})$

Hence the efficiency with which energy is transferred from a transient blast of the solar wind to the terrestrial magnetosphere is only about 0.1%. The interesting point is that though this conversion factor appears to be very low, it still is much higher than the value predicted by any of the available theories. As a result the exact mechanism by which part of the energy of the solar blast is transferred to the geomagnetic cavity remains still unclear.

In Section 4.4 we had seen that the total energy produced during a flare was approximately 10^{31} ergs. For a typical duration of $\sim 10^3$ sec, the average energy output is $W \simeq 10^{28}$ erg/sec. The plasma cloud ejected by the flare has a certain collimation since we usually do not detect the flares which occur in the back hemisphere of the sun. The solid angle in which the cloud is ejected is probably close to 1/3 of the entire sphere and therefore the corresponding area ΩR^2 at $R = 1$ A.U. will be of the order of 10^{27} cm^2. Hence the energy flux of the solar blast at the orbit of the earth is,

$$F = \frac{W}{\Omega R^2} \simeq \frac{10^{28}}{10^{27}} \simeq 10 \text{ erg cm}^{-2} \text{ sec}^{-1} \qquad (6.1\text{-}7)$$

which is essentially the same with the value we obtained in (6.1-4).

6.2   Geomagnetic Storms and Ring Currents

The eruption of a solar flare results usually in the ejection of a plasma cloud into the interplanetary space. The radial velocity of this cloud is typically 500–1000 km/sec, which is 1.5 to 3 times the velocity of the quiet solar wind, and the plasma cloud reaches the earth in 1.5 to 3 days. It is of interest to mention that Obayashi (1962) has found that when several geomagnetic storms occur in a span of a few days, the plasma cloud that causes the *leading storm* takes almost 50% longer to arrive than the clouds of the *following storms*. This might be due to the streamlining of the interplanetary magnetic field by the cloud of the leading storm, which in a sense paves the way for the plasma clouds to follow.

The interception of the plasma cloud by the terrestrial magnetosphere signals the beginning of a magnetic storm on earth. Figure 6.2-I shows the time history of a typical geomagnetic storm in terms of the changes in the horizontal component of the earth's magnetic field observed from a station at low latitudes. The preference to the horizontal component of the magnetic field and to a low latitude station comes from the fact that they show the most consistent variations during a geomagnetic storm. As seen from Figure 6.2-I, the storm starts with a sudden increase of the magnetic field by 20–50 γ. This is called *sudden storm commencement* and is often denoted with the letters SSC. It should be mentioned, however, that the *storm commencement* (SC) is not always a sudden one. Some storms start more gradually and then we speak of a *gradual storm commencement* (GSC). The increase of the magnetic field lasts from one hour to several hours and represents the *initial phase* of the magnetic storm. This increase is produced by the compression of the magnetosphere under the pressure of the plasma cloud. Some people denote the magnetic field which produces the initial phase with the letters DCF, which are the initials of the words *disturbance corpuscular flux.*

Following this brief increase, the magnetic field begins to decrease, and reaches a value of 100–300 γ below the quiet-time level in about 12 hours. After the maximum decrease, the magnetic field starts a slow recovery to pre-storm conditions, which can last for one to three days. The period when the magnetic field is below the normal level is called the *main phase* of the magnetic storm. Some people denote the magnetic field responsible for the main phase of the storm with the letters DR which are the initials of the words *disturbance ring.* The decrease of the magnetic field is due to the build-up of a new radiation belt of low energy (200 eV $< E <$ 50 keV) charged particles around the earth, which most probably enter into the outer Van Allen belt from the magnetic tail of the earth. In the equatorial plane this zone peaks at a distance of approximately 4 earth radii and produces an electric current around the earth which is called the *ring current.*

The recovery period of a geomagnetic storm shows often two distinct stages with different rates of recovery. The first one follows the maximum decrease and lasts for about half a day. The second stage has a much

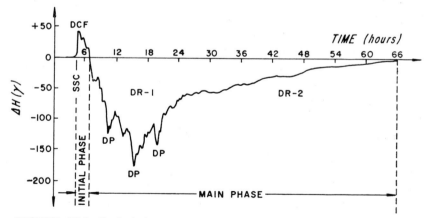

FIGURE 6.2-I   Typical changes (in gammas) of the horizontal component of the earth's magnetic field at mid-latitudes during a moderately strong magnetic storm

slower recovery rate and can last for 1–3 days before the magnetic field of the earth reaches its pre-storm level. As a result, the DR is believed to consist of two fields, called DR1 and DR2. The DR1 is the field due to the main ring current mentioned above, while the DR2 is attributed to a secondary ring current, either further out in the magnetosphere or in the geomagnetic tail.

Intermittently during the main phase we can observe several impulsive fluctuations of the magnetic field. These disturbances, which last only 1–3 hours, are due to currents flowing mostly in the ionosphere of the polar region to which they are communicated from the magnetosphere. The magnetic field generated by these polar currents is often denoted with the letters DP, which are the initials of the words *disturbance polar*. Though these disturbances are most intense in the polar region, DP's can be observed also at low latitudes. They appear as irregular changes which are superimposed on the main phase decrease (Figure 6.2-I). The most familiar DP current is the *auroral electrojet* which circles the polar region along the auroral belt and is associated with *polar substorms* which last also only 1–3 hours. In spite of their brief duration, they produce occasionally changes of several thousand gammas in the magnetic field of the auroral region.

It is believed that the main phase of a geomagnetic storm is actually a sequence of such intermittently occurring impulsive substorms. If these substorms occur frequently enough (Figure 6.2-IIb), the low energy charged particles injected into the outer belt will accumulate and will form a strong ring current at a distance of approximately 4 earth radii (Cahill, 1966). Direct particle measurements in the magnetosphere during magnetic storms (Frank, 1967 and 1970; Vasiliunas, 1968) show an increase by several orders of magnitude in the densities of these low energy particles at distances of 4–5 earth radii near the equatorial plane. If, on the other hand the substorms are separated by longer time intervals (more than about 12 hours), the ring current will decay after each substorm (Figure 6.2-IIa) because most of the injected particles will disappear during this long time interval through charge-exchange collisions with neutral atmospheric particles. As a result the substorms will not be able to build a lasting ring current and the geomagnetic disturbance will have the form shown in Figure 6.2-IIa.

The auroral electrojet, is not the only current known to flow in the terrestrial ionosphere. Actually there is a system of currents flowing over the entire earth produced by the day-night assymmetry in the temperature of the neutral atmosphere, and in the electron density and temperature of the terrestrial ionosphere. This system of global currents is called the *Sq-current* and is responsible for the quiet-day diurnal variation of the earth's

11*

magnetic field. The Sq-current flows mainly in the E-layer of the ionosphere (100–125 km) forming essentially symmetric loops about the geomagnetic equator. The current flow is stronger in the sunlit hemisphere and reaches a peak around midday in a band extending to about 2° on either side of the geomagnetic equator. The strong current in this band is called the *equatorial electrojet* because it reminds one of the jet stream that flows in the stratosphere. The equatorial electrojet can at times change the ground magnetic field by 1–2%.

The magnetic field produced by a current flowing in a wire is shown in Figure 6.2-III. The direction of the current is by definition the direction in which the positive charges would move, and the convention for the magnetic field is the same used in the case of the earth's magnetic field where the field lines exit from the south hemisphere and re-enter in the north hemisphere. The auroral electrojet runs in the westward direction in the auroral zone around the polar cap. As a result the magnetic field it produces adds to the magnetic field of the earth in the region around the pole, but opposes the terrestrial magnetic field outside the polar cap. An area showing an increase in the magnetic field is called a *positive bay* and an area showing a decrease is called a *negative bay*. The magnetic disturbance in the polar regions is a more complex arrangement of positive and negative bays than

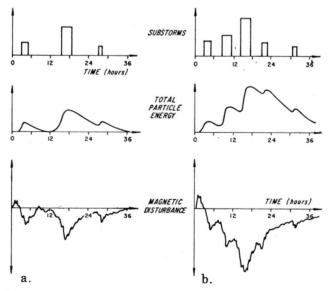

FIGURE 6.2-II   The build-up of the magnetic disturbance and the total particle energy of the ring current in the case of: (a) few substorms separated by longer time intervals, and (b) larger number of sub-storms separated by shorter time intervals

the auroral electrojet alone could account for. For this reason we were forced to introduce an additional current system which is denoted as DP-2, to distinguish it from the auroral electrojet which in this scheme is denoted DP-1. The DP-2 current system flows directly over the polar cap and closes its electric loops by forming two vortices, one in the morning side and one in the evening side of the polar cap.

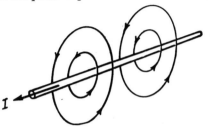

FIGURE 6.2-III   The magnetic field lines produced by a current (by convention, the flow of positive charges) passing through a wire

Until recently these current systems in the ionosphere were inferred only from the magnetic field observations on the ground. For this reason they were referred to as the *equivalent current systems*. A few years ago, however, a group of German scientists from the Max Planck Institute (Föppl *et al.*, 1967) developed a method which allows the direct measurement of the electric fields in the ionosphere which produce these currents.

Barium vapor, released from a rocket at an altitude of approximately 150 km, becomes partially ionized and produces a visible ion cloud. The electric fields in the ionosphere induce a drift velocity to this luminous cloud which can be measured photographically. This drift velocity is proportional to the electric field and with this simple technique we can measure directly the electric fields in the ionosphere. Measurements of electric fields in the ionosphere have also been made recently with sounding rockets and satellites using long antennas, but the barium vapor method has the advantage that allows us to conduct our observations over a much larger area and for considerably longer periods of time.

We have mentioned already that the main phase of the magnetic storm is due to a ring current which is produced by the injection of a large number of low energy charged particles at a distance of 4–5 R_0. It should be emphasized that the energy of these particles (in the keV range) is much higher than the energy of the ionospheric plasma (~ 0.3 eV). To understand the generation of the ring current let us examine a small section in the equatorial plane of the belt, which these newly injected particles form around the earth. Let R_b be the distance at which the particle density of the belt reaches a maximum (Figure 6.2-IV).

Since this is an equatorial cross-section, the magnetic field is normal to the plane of the diagram. The orbits of the charged particles which are essentially normal to the magnetic field, are in the plane of the diagram. Let us first consider only particles with positive charge (protons) so that all orbits will have the same sense of rotation. As seen from Figure 6.2-IV, all the radial components of the little loop currents cancel out (equal number of arrows in the $+R$ and the $-R$ direction). The tangential components, however, cancel only at $R = R_b$ while on either side of R_b, due to the decreases in the particle density, they combine to form two currents that run in opposite directions.

Particles with negative charge (electrons) have the same spatial distribution but they will spiral around the magnetic field lines in the opposite sense. As a result the sum of all the little arrows will yield a vector in the opposite direction. A current, however, can be defined either as positive charges moving in one direction or as negative charges moving in the opposite direction. As a result, the current produced by the motion of the protons and the current produced by the motion of the electrons in the opposite direction add together to form a single current. In the case of the earth the result is an eastward current (electrons moving westward

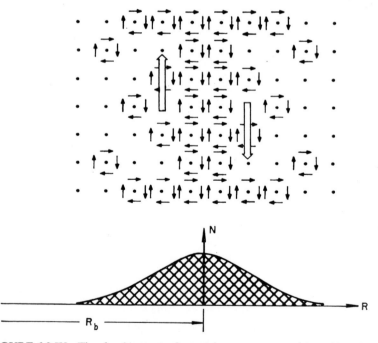

FIGURE 6.2-IV The development of opposite currents on either side of the maximum of the belt, due to decreasing particle densities on either side of the peak

and protons moving eastward) inside the density maximum ($R < R_b$) and a westward current (electrons moving eastward and protons moving westward) outside the density maximum ($R > R_b$).

In Section 3.3 we have seen that as the trapped particles bounce between the two conjugate hemispheres they also are forced to move in the azimuthal direction around the earth, the electrons drifting toward the east and the protons toward the west. This is simply a westward current which opposes and nearly cancels the eastward current inside the maximum of the belt and intensifies the westward current outside the maximum of the belt. So essentially we are left with a westward ring current around the earth (Figure 6.2-V). Using the convention shown in Figure 6.2-II, we can readily see that the magnetic field due to the ring current will oppose the earth's magnetic field inside the space encircled by the belt of the ring current (Figure 6.2-V).

From the above discussion it is obvious that the intensity of the ring current will be proportional to the total number of particles in the belt which determines the total available charge, and to their kinetic energy which determines their azimuthal velocity. On the basis of these theoretical deductions, Dessler and Parker (1959) derived the semiempirical expression,

$$\Delta H_0 = -2.6 \times 10^{-21} E \qquad (6.2\text{-}1)$$

where ΔH_0 (in gammas) is the decrease of the equatorial magnetic field on the surface of the earth, and E (in ergs) is the total energy of the low energy charged particles that produce the ring current. Taking the center of the belt at approximately $4R_0$ and its cross-section about $4R_0 \times 4R_0$, we find that the volume of the entire belt is equal to $\sim 10^{29}$ cm^3. Taking the typical particle energy to be ~ 5 keV, and the average particle density equal to ~ 100 protons/cm^3, we find that the average energy density is

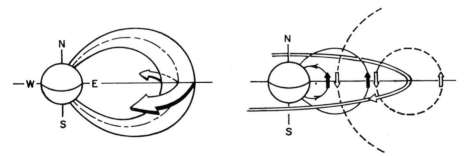

FIGURE 6.2-V   The strong westward current which develops outside the maximum of the belt. The magnetic field produced by this current opposes the terrestrial magnetic field in the region between the earth and the ring current

$\sim 8 \times 10^{-7}$ ergs/cm³. Hence the total particle energy in the zone of the ring current is,

$$E \simeq 8 \times 10^{-7} \times 10^{29} \simeq 8 \times 10^{22} \text{ ergs} \qquad (6.2\text{-}2)$$

and using this value of E in (6.2-1) we get,

$$\Delta H_0 \sim -200\gamma \qquad (6.2\text{-}3)$$

which is in agreement with the field decreases observed during a relatively strong magnetic storm.

Measurements of magnetic fields and low energy particles in the magneto-sphere during geomagnetic storms confirm in essence the above computations. These in situ measurements suggest that the low energy particles are injected into the outer radiation belt from the dark side of the earth and probably from the plasma sheet of the magnetic tail which has been observed to advance much closer to the plasmapause during magnetic disturbances (Vasyliunas, 1968). It has been found also that the energy of the injected particles rests $\sim 80\%$ with the protons and only $\sim 20\%$ with the electrons. In accordance with the westward current we discussed, the protons from the night side of the earth drift west toward the sunset region, while the electrons drift east toward the sunrise region. As a result, in the early hours of the storm there is a considerable anisotropy around the earth (Frank, 1970) and the maximum *inflation* of the magnetic field appears in the evening quadrant. This effect is verified by simultaneous measurements at different geomagnetic stations around the globe. The word "inflation" is often used to describe a decrease of the magnetic field, because by inflating the earth's magnetic field, i.e. by distributing the same number of field lines inside a larger volume, we decrease the density of the field lines and therefore the intensity of the magnetic field. A compression of the magnetosphere, on the other hand, produces an increase in the ground magnetic field, a phenomenon which actually occurs during the initial phase of the magnetic storm.

After a period of a few hours to one day, the drifting of the particles around the earth evens out the particle density around the belt and the magnetic disturbances recorded in all quadrants become very similar. In the case of a very weak storm a fully symmetric current might never be established, because a weak storm (2.6-1) starts probably with a smaller total number of particles, which in addition have a lower average energy. As seen, however, from (3.3-16) and (3.3-21), particles of lower energy at the same L-shell have to make a larger number of mirror reflections during a given longitude drift. As a result, it is very likely that they will be dissipated through charge exchange with the neutral atmosphere at the mirroring points before they have a chance to establish a uniform ring current. Note that a 5 keV proton at $L = 4$, will need about 36 hours to complete one revolution around the earth and will make more than 1,000 mirror oscillations during this time interval.

It has been observed that the establishment of a symmetric ring current occurs faster in the bigger storms. This is probably because the bigger storms have a larger fraction of particles in the higher energy bracket which drift with higher velocities around the earth and thus equalize faster the particle density around the belt. There seems to be some evidence also that the ring current of the more intense storms is established closer to the earth. This would explain the faster decay of the larger decreases of the magnetic field because, when the belt is set up closer to the earth, the particles of the ring current mirror more times during a given longitude drift and on the average penetrate deeper into the neutral atmosphere at the mirror points. The result of course is that they are lost faster and the strong ring current decays in a shorter time interval.

We have now reached the point where the magnetic disturbances measured on the ground at all latitudes and all local times, fit nicely together into a consistent pattern with our satellite observations of particle and field changes in the magnetosphere. From all these studies we have finally assembled the basic model for the entire magnetic storm, but the picture is not yet totally clear and there are still many details that need further study and more observations. The two most important problems that remain still unclear are, the process through which the charged particles from the sun find their way into the magnetosphere of the earth, and the mechanism by which these particles are heated up (accelerated) to the energies at which they appear in the radiation belts and in the auroral zone.

6.3 Galactic and Solar Cosmic Rays

The earth is continuously bombarded by energetic charged particles of cosmic (galactic and extragalactic) origin. These particles are called *cosmic rays*. Their origin and acceleration mechanism remain still uncertain, though the *pulsars* might provide finally the answers to these questions. Occasionally, the earth is bombarded also by energetic particles from the sun which are called *solar cosmic rays*. These are produced in large solar flares but their acceleration and storage are not very well understood either. Since the cosmic rays are high energy charged particles, their trajectories toward the earth will be modified by the magnetic fields that surround our planet. Actually these fields might even prevent some of the cosmic ray particles from reaching the surface of the earth.

Given the direction of arrival, the geomagnetic latitude λ, the mass m, and the charge Ze of a cosmic ray particle, one can compute the *critical momentum* p_c that the particle must have in order to overcome the deflection of the earth's magnetic field. For vertical incidence in a dipole magnetic field the minimal momentum required is,

$$p_c = mV = \frac{ZeM}{4R_0^2 c} \cos^4 \lambda \tag{6.3-1}$$

where $M = 8.05 \times 10^{25}$ Gauss cm^3 is the dipole moment of the earth's magnetic field, and R_0 is the radius of the earth. Equation (6.3-1) shows that the critical momentum required decreases as the latitude increases. As a result we can expect to observe a much higher flux of cosmic rays in the polar region than in any other part of the earth. Equation (6.3-1) is valid also for relativistic particles for which the relativistic momentum p, can be expressed in terms of the total energy E_T and the kinetic energy E_k through the following relations,

$$(pc)^2 = (mVc)^2 = E_T^2 - (m_0 c^2)^2 = E_k^2 + 2E_k m_0 c^2 \tag{6.3-2}$$

Particle energies are often expressed in electron volts ($1\,\text{eV} = 1.6 \times 10^{-12}$ ergs), and since pc represents a quantity of energy, it can be expressed in eV too. As a result, pc/e can be expressed in volts. For cosmic ray particles, it has been found convenient to use a parameter called *rigidity P*, which is defined by the following relation,

$$P = \frac{pc}{(Ze)} \tag{6.3-3}$$

and describes the penetrating ability of charged particles in a magnetic field. From (6.3-1) and (6.3-3) we find that the *cut-off rigidity* of particles arriving vertically at a geomagnetic latitude λ is,

$$P_c = \frac{p_c c}{(Ze)} = \frac{M}{4R_0^2} \cos^4 \lambda = 14.9 \cos^4 \lambda \; \text{GV} \tag{6.3-4}$$

where $1\,\text{GV} = 10^9$ volts. The rigidity and the kinetic energy of a particle can be related through (6.3-4) and (6.3-2). For a vertically incident proton, for example, at the equator ($\lambda = 0$) we have $P_c = 14.9\,\text{GV}$ and $E_k = 14.0\,\text{GeV}$, while at $\lambda = 60°$ we find $P_c = 931\,\text{MV}$ and $E_k = 384\,\text{MeV}$.

The cosmic ray particles incident upon the top of the atmosphere are called *primary cosmic rays*. These particles interact with the nuclei of the atmospheric molecules and produce *secondary cosmic rays*. These in turn interact with other nuclei producing more secondary particles, and so on.

Thus a primary cosmic ray causes a cascade of secondary cosmic rays which is called a *cosmic ray shower*. The production of secondary particles becomes significant at about 55 km and reaches a maximum at an altitude of about 20 km. Below this height and up to the ground, the intensity of the secondaries decreases as the particles lose energy through collisions until most of them either decay or are absorbed. Only primaries with $E_k > 500$ MeV have a fair chance of reaching the ground, but the primary flux impinging on the top of the atmosphere with $E_k > 500$ MeV is only about $0.2 \, \text{cm}^{-2} \, \text{sec}^{-1} \, \text{sr}^{-1}$ for protons and $0.03 \, \text{cm}^{-2} \, \text{sec}^{-1} \, \text{sr}^{-1}$ for alpha particles.

Neutrons and μ-mesons are the secondaries usually measured on the surface of the earth to obtain information on the flux and energy spectrum of the primaries. Ionization chambers are used for the counting of energetic μ-mesons. Neutron monitors are in general more suitable for the study of variations in the primary flux because they are more sensitive indicators of primary radiation with energies in the range from 0.5 to 10 GeV. It should be noted that all ground measurements must be corrected for the time variations of the atmospheric thickness (pressure) and the variations of the atmospheric depth with latitude and altitude. Thus in order to compare cosmic ray measurements from around the world we must first normalize them to a standard atmosphere.

Observations over several decades have revealed that the cosmic ray flux is modulated by the 11-year solar cycle, reaching a maximum during the quiet period of the solar cycle and a minimum near the peak of solar activity. The explanation of this effect is that during the active period of the solar cycle, more flare events pull more of the solar magnetic field into the interplanetary space thus providing additional shielding against the cosmic rays. The low energy component of the cosmic rays in the component showing the highest solar cycle modulation, while particles with rigidities higher than 15 GV seem to remain relatively unaffected.

There is also a 27-day modulation of the cosmic ray flux because the interplanetary magnetic field co-rotates with the active regions on the sun from which it stems. There are also occasional sudden decreases in the cosmic ray flux with a recovery period of several days. This phenomenon is called a *Forbush decrease* after Dr. Scott E. Forbush who discovered it. These decreases range from a few per cent to more than 25 per cent of the integrated relativistic spectrum of the cosmic rays and they extend to particles with rigidities beyond 50 GV.

The two models which have been proposed to explain the Forbush decreases are shown in Figure 6.3-I. According to the first model, which was proposed by T. Gold (1959), a plasma cloud ejected by a large flare event can draw out the magnetic field of the sun forming a large magnetic "bottle". Under certain geometric conditions, this "bottle" can engulf the earth and provide the additional shielding which explains the Forbush decrease in the cosmic ray flux.

According to the other model, which was proposed by E. Parker (1961), following the eruption of a large flare a blast wave develops which propagates in the interplanetary space causing a sharp local increase of the interplanetary magnetic field. This shell of higher field intensity provides the extra protection against the cosmic rays. Several space-probes, such as Mariner 2 and Explorer 34, have observed this type of an abrupt increase of the interplanetary magnetic field which accompanies a propagating shock wave. An enhancement in the cosmic ray flux, which is occasionally observed after a large flare (*proton event*) and before the onset of the Forbush decrease, is produced by solar cosmic rays which are emitted during certain flare events. A brief increase which sometimes occurs before the beginning of a Forbush decrease is probably related to the arrival of the shock wave from the sun which in this case contains a relatively large number of semi-trapped solar cosmic rays.

In addition to the 11-year cycle and the 27-day variations, the cosmic ray flux displays also a *diurnal anisotropy* with a maximum near sunset and a minimum near sunrise. The magnitude of the anisotropy is latitude dependent and reaches its peak at the equator where the amplitude variation is 0.8% from sunrise to sunset or half of it, i.e., 0.4% from either extreme

GOLD'S MODEL

PARKER'S MODEL

FIGURE 6.3-I   Gold's and Parker's models of the enhanced interplanetary magnetic field which produces the Forbush decrease of the cosmic ray flux on earth

to noon or to midnight. As the earth moves in its orbit around the sun, sunset and therefore the maximum of the cosmic ray flux occurs in the trailing sector of the earth. This observation gives a good clue for the cause of the diurnal anisotropy.

The cosmic rays, in addition to their random velocity W, which is nearly equal to the velocity of light ($W \simeq c = 3 \times 10^{10}$ cm/sec), have also a common drift velocity V because they follow the interplanetary magnetic field as it co-rotates with the sun. The tangential velocity at 1 A.U. for a rigid rotation with the sun is $V \simeq 4 \times 10^7$ cm/sec. The earth, on the other hand, moves around the sun with an orbital velocity of approximately 3×10^6 cm/sec which is negligible comparing to V. As a result a cloud of cosmic rays overtakes continuously the earth which in this case can be considered as practically stationary. An observer being overtaken by this cloud will think that the cosmic ray particles have a velocity,

$$W' = W + V = W\left(1 + \frac{V}{W}\right) \tag{6.3-5}$$

To the approaching particles, on the other hand, the acceptance cone angle θ of the cosmic ray counters will appear to be,

$$\theta' = \theta\left(1 + \frac{V}{W}\right) \tag{6.3-6}$$

and the acceptance solid angle of the counters,

$$\Omega' = (\theta')^2 = \theta^2\left(1 + \frac{V}{W}\right)^2 = \Omega\left(1 + \frac{V}{W}\right)^2 \tag{6.3-7}$$

Since the counting rate of cosmic rays is proportional to the velocity of the particles W' and the solid angle of the counters Ω', the cosmic ray flux measured F' will be,

$$F' = F\left(1 + \frac{V}{W}\right)^3 \simeq F\left(1 + \frac{3V}{W}\right) \tag{6.3-8}$$

where F is the corresponding flux near the noon or the midnight meridian where the cosmic ray cloud moves tangentially to the ground and its motion does not affect the counting rate. From (6.3-8) it follows that,

$$\frac{(F' - F)}{F} = \frac{3V}{W} = \frac{3 \times 4 \times 10^7}{3 \times 10^{10}} = 0.4\% \tag{6.3-9}$$

This theoretical result (Parker, 1964) is in good agreement with the existing experimental observations and confirms the hypothesis that the cosmic rays partake in the co-rotation of the interplanetary magnetic field with the sun. Annual means of the diurnal anisotropy, with a maximum as we said near 18:00 hours, show variations that are well correlated with magnetic activity and the solar cycle (Forbush, 1969). The annual maximum occurs near sunspot maximum and is 0.4–0.5% against 0.3–0.4% for the minimum which occurs at the solar minimum. The difference is due to the enhancement of the interplanetary magnetic field which allows the trapping of cosmic rays of higher energies and thus increases the fraction of the cosmic rays which co-rotates with the interplanetary magnetic field.

Cosmic rays which manage to penetrate into the atmosphere of the earth are responsible for maintaining the present level of certain radioactive isotopes which otherwise would not be there because of their short half-life periods. A typical example is C^{14} which, with a half-life of only 5,000 years, is used extensively in radioactive dating of archaeological findings. The C^{14} dating method gives results that are in very good agreement with other archaeological dating methods, which implies that during the past 30,000 years the cosmic ray flux has remained constant to better than a few per cent.

The high fluxes of the solar cosmic rays emitted by large flare events can produce severe biological effects to space travellers. As seen from

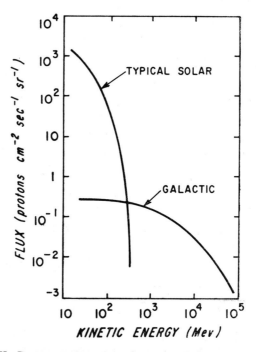

FIGURE 6.3-II   Proton energy spectra from a typical solar proton event and of the galactic cosmic ray flux near solar minimum

Figure 6.3-II, the energy spectrum of a typical solar event does not extend to as high energies as does the spectrum of the galactic cosmic rays. Around 100 MeV, however, solar cosmic ray fluxes can exceed the galactic flux by several orders of magnitude and therefore they can be very dangerous for space travellers. The *rad* is the unit most frequently used for the amount of radiation absorbed. The definition of this unit is that 1 rad = 100 ergs of radiation absorbed per gram of absorbing material. The biological effects depend on the total dose absorbed and the rate at which it was received.

It is estimated that an unprotected person exposed in space to a large solar event can receive an integrated dose of approximately 1,000 rads in a few hours. This dose is far beyond any tolerable limit. For this reason, a large research effort has been undertaken to evaluate the biological effects of radiation in space and to compute the most economic, but still effective, shielding for manned spacecrafts.

Research is also directed toward the development of a reliable forecasting system for large solar events. There are indications for example that the development of the penumbra in the early stages of the different sunspots can provide valuable information in this direction. Presumably the development of the penumbra is directly related to the magnetic fields present, and as we know the strength and the configuration of the magnetic fields in the sunspot region are the decisive factors which determine the occurrence and the magnitude of solar flares. Solar cosmic rays are in general much more dangerous than the x-rays emitted during a solar flare. The basic reason is that while x-ray bursts usually last less than one hour, solar cosmic ray events can last for more than one day. As a result they are capable of delivering a much higher dose of radiation to an unprotected space traveller.

Cosmic rays, through pair production and collisions produce a certain amount of ionization in the atmosphere. Their contribution, as a matter of fact, in the formation of the polar ionosphere is quite important because the cosmic ray flux is much higher over the polar caps while the solar ultraviolet and x-ray flux in these regions is quite low. The ionization rate due to the cosmic rays can be as high as several hundred ions per second per cm^3. Most of it, however, is produced at altitudes near 30 km where the density of the atmosphere is sufficiently high to cause almost instant recombination.

A large flux of solar cosmic rays can cause in the polar regions a very significant increase in the ionization of the lower ionosphere and especially of the D-layer. As we have seen in Section 2.4, an enhanced D-region will produce a strong attentuation to the radio waves, and especially to the low frequency ones that pass through it. The strong radio absorption produced by the sudden influx of solar cosmic rays in the polar regions is called *polar cap absorption* (PCA). The level of absorption is usually measured with riometers which monitor continuously the intensity of the galactic radio-noise background at different frequencies.

Most solar cosmic rays have energies from a few MeV to a few hundred MeV and therefore do not have the rigidity to pass through the earth's magnetic field outside the polar caps. For this reason a PCA event remains essentially the only practical way to detect solar cosmic rays from the ground. PCA events are very seldom associated with flares of importance less than 2, and even among the larger ones only about 20% of the $2+$, 3, and $3+$ flares produce detectable PCA's. The reasons might be traced to the nature of the particular flare as well as to its position on the solar disc and the morphology of the interplanetary magnetic field at the time of the event. Flares in the western side of the sun, for example, are more likely to produce high intensity solar cosmic ray events on earth because of the spiral configuration of the interplanetary magnetic field. During solar maxima we have about one PCA event per month, while during solar minima we can anticipate at most one or two events per year. These figures depend of course also on the sensitivity limit of the riometers used in these observations.

Since few solar cosmic rays have energies about 500 MeV, we cannot expect to detect them on the surface of the earth because neither their primaries nor their secondaries can reach the ground. As a result solar cosmic rays can be observed directly only with counters flown on satellites. Such observations are now part of a solar patrol program. Occasionally some very large flares produce a significant flux of solar cosmic rays with energies higher than 500 MeV which can be detected with neutron monitors from the ground. These rather rare events are called *sea level events* or *ground events*. Around the solar maximum they occur at the rate of 2–3 per year, but they are practically absent around the solar minimum.

Solar cosmic ray events are also called *solar proton events*, or simply *proton events*, though solar cosmic rays usually include also a significant fraction of alpha particles. Protons and alpha particles tend to have similar rigidity spectra but their ratio varies considerably from one event to another. This might possibly be due to nuclear reactions in the core of the flare, but most likely it is related to the acceleration process and the trapping of the high energy particles in the flare region. In any case, it remains still one of the most puzzling aspects of solar cosmic ray physics.

The beginning of a PCA event, which signals the arrival of the first solar cosmic rays, occurs approximately one hour after the detection of the x-ray burst or the microwave burst from the flare event. This time lag, which can actually vary from 20 minutes to several hours, should not be attributed only to the velocity of the solar cosmic rays. It can represent also the time interval required for the acceleration of these particles in the region of the flare, or it can represent the time interval which passed before these particles were able to escape from the strong solar magnetic field over the flare region. Finally it might also be an indication of the curved paths due to the interplanetary magnetic field which these particles had to follow in their journey from the sun to the earth, or the propagation difficulties which they encountered due to a tangling of the magnetic field lines in the interplanetary space.

The flux of the solar cosmic rays in the vicinity of the earth increases rapidly following the arrival of the first particles. The build up stage of a PCA event lasts from 20 minutes to a few hours and seems to be longer for events of lower energy. The build up stage is followed by a transition period which lasts for a few hours and is characterized by large fluctuations in the particle flux. After this the decay stage starts which can last from half a day to a few days. All in all, after some large flares solar cosmic rays continue to bombard the earth for one or more days. This requires either a continuous production of cosmic rays in the flare region, which seems very unlikely, or some appropriate storage mechanism for the solar cosmic rays that were produced during the flare event.

Solar proton events are well correlated with type-IV solar radio bursts, which can last for many hours and represent the synchrotron radiation emitted by energetic electrons trapped in the solar magnetic fields above the active region. These magnetic fields could provide also a storage region for the solar cosmic rays, which can then be released continuously as the magnetic fields decay. Another possibility is that solar cosmic rays remain partially trapped in the magnetic fields frozen in the plasma cloud which is ejected from the flare region. Since they are only partially confined, a fraction of the solar cosmic rays escapes continuously as the plasma cloud advances through the interplanetary space. This theory is supported also by the fact that proton events occur occasionally on the earth without a previous flare on the sun. In these cases a large flare could have occurred in the back side of the sun and when the ejected plasma cloud cleared the disc of the sun, some of the solar cosmic rays which were partially confined in the cloud, were able to find their way to the earth along some favorable paths of the interplanetary magnetic field.

The importance of the alignment of the interplanetary magnetic field becomes clear when two large flares occur a few days apart in the same western region of the sun. In these cases the solar cosmic rays from the second flare find a nicely collimated path to earth and produce a much more intense solar proton event. It is of interest to note also that field aligned streams of solar cosmic rays can persist for weeks or months co-rotating with the interplanetary magnetic field. As a result, 27 days after a strong proton event we might observe again on the earth a weak influx of solar cosmic rays.

Much more work is still needed to fully understand all the aspects (acceleration, energy spectra, atomic ratios, storage mechanisms, etc.) of solar cosmic ray physics. But this is also what makes them one of the most interesting topics in the entire field of space physics and space astronomy.

6.4 Auroras

An *aurora* is one of the most spectacular phenomena of nature. Reports of this impressive visual effect exist for at least 2,000 years, and careful aurora observations exist for more than 300 years. The name *aurora borealis* was introduced by the French astronomer Gassendi when he described the large aurora display of September 12 1621. It is now used for the auroras of the northern hemisphere. Auroras in the southern polar regions were first reported by Captain Cook in 1773, and were given the name *aurora australis*. When no hemisphere is specified the aurora is called *aurora polaris*. In 1722 Graham in London, and independently in 1741 Celsius in Sweden, discovered that auroras are closely associated with geomagnetic disturbances.

Plots of *auroral isophotes*, i.e., of the average number of auroras seen at different locations per year, show that aurora sightings reach a maximum around 65°–70° geomagnetic latitude. This is called the *auroral zone* and its existence has been known for at least hundred years. Equally old is our knowledge that the number of auroras is well correlated with the sun-spot cycle and it is of interest to add that auroras were one of the most important subjects of the first *international polar year* in 1882–1883.

Inspite of all this past work, the correct geometry of the auroral region was discovered only in the last few years through the extensive use of all-sky cameras and fast jet planes. These modern techniques made it possible to study the development of auroras all around the polar cap throughout the entire night and day period. These studies have shown that the maximum of the diurnal auroral activity occurs in an oval band (and not in a circular zone) around the magnetic pole. This oval region (Figure 6.4-I), which was named the *auroral oval*, is actually the intersection of the outer shell of trapped radiation with the terrestrial atmosphere. It corresponds to $L = 6$–7. The assymmetry of the oval, as seen from Figure 6.4-II, is due to the distortion of the earth's magnetic field by the solar wind.

The auroral oval crosses the noon meridian at ~76° geomagnetic latitude, and the midnight meridian at ~67°. Because, as we will see auroras are at their brightest near the midnight meridian, most causal auroral sightings were made around local midnight. This explains why past workers had reached the erroneous conclusion that the auroral zone, i.e., the zone of maximum auroral activity is a symmetric circle around the geomagnetic pole at about 67° geomagnetic latitude. As seen from Figure 6.4-I, the auroral zone (dotted circles) is simply the trace of the midnight sector of the auroral oval as the earth rotates under the oval.

With increasing magnetic activity, the auroral oval becomes wider and its lower boundary advances significantly toward the equator. During

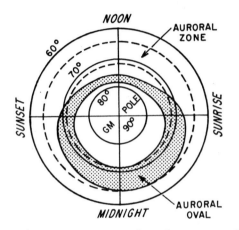

FIGURE 6.4-I   A polar diagram, centered at the geomagnetic pole, showing the auroral oval and the auroral zone

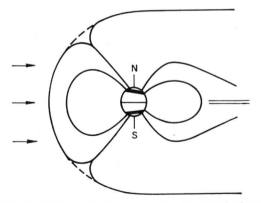

FIGURE 6.4-II   The higher latitudes on the dayside and the lower latitudes on the nightside of the auroral oval are due to the deformation of the earth's magnetic field by the solar wind

12*

intense magnetic storms it can reach as low as 50° geomagnetic latitude in the midnight sector. The shape of the auroral oval undergoes also some small diurnal and seasonal variations because the axis of the terrestrial magnetic field is tilted by about 11° to the rotational axis of the earth, which in turn is tilted by about 23° to the normal to the ecliptic, which is the plane of the earth's orbit around the sun.

Auroras, in spite of their many forms and colors, belong essentially in one of the following three categories:

Quiet arcs   These are thin luminous sheets of rather uniform brightness extending for several thousand kilometers around the auroral oval. Their thickness is about 3 km and they are inclined to the vertical in the direction of the dip of the magnetic field. They display only very slow motions.

Rayed arcs   These are striated along the magnetic field lines and they are wavy or folded. They often display rapid motions giving the impression of huge draperies waving in the sky. Their thickness is extremely small, usually only of the order of 300 m, and they are usually seen in the pre-midnight sector.

Diffuse patches   These are luminous regions in the sky of no specific shape or border. Their luminosity is often seen to pulsate, and they are usually seen in the post-midnight sector.

The quiet arcs are essentially a continuous polar phenomenon, but they do have periods of higher or lower intensity. Periods of intense auroral activity are called *auroral substorms*. These usually start in the midnight sector of the auroral oval and spread rapidly over the entire oval region. Auroral substorms last 1 to 3 hours, and we can have several of them per day during the course of a large geomagnetic storm. They are usually associated with polar magnetic substorms which are due to the intense currents (auroral electrojet) that are generated in the polar ionosphere. Polar substorms last also 1–3 hours and represent, as we have seen, the elementary units of a geomagnetic storm.

The cycle of an auroral substorm starts usually in the midnight region of the oval where a sudden ($t = 0$–5 min) brightening of one of the quiet arcs takes place. This is followed by a rapid motion of the active auroras toward the poles. This is called a *poleward surge* and produces an *auroral bulge* around the midnight sector ($t = 5$–10 min). The *break-up* of the auroras takes place within the auroral bulge which, in addition to the expansion toward the pole, also expands toward the east, the west, and the equator ($t = 10$–30 min). The expansion toward the west takes the form of a large fold (rayed arcs) travelling rapidly with a velocity of about 1 km/sec toward the dusk sector. This is called the *westward surge* and can travel for several thousand kilometers. The expansion toward the east results in the disintegration of the arcs into diffuse patches which drift toward the dawn sector with a velocity of about 300 m/sec.

After reaching its maximum size ($t = 30$–60 min), the bulge starts to contract and the auroral oval returns to its prestorm conditions in another hour or so. During the recovery stage the westward surge continues to travel into the evening sector where it ultimately degenerates into irregular bands. In the morning sector, on the other hand, the eastward drifting patches remain visible up to the very end of the recovery stage spreading also over a large latitude area. The basic steps of an auroral substorm are outlined in Figure 6.4-III. From the above, it is quite clear that most of the aurora activity occurs near the midnight sector of the auroral oval which, as we have seen, explains why past workers had mistaken the auroral oval as a circular auroral zone.

Field aligned bundles of ionization, often associated with optical auroras, can be detected with radar echoes and constitute the so-called *radio aurora*. The relation of the radio auroras to the optical auroras is still somewhat unclear because they do not show a one to one correspondence.

Auroras are produced by energetic electrons (~ 10 keV) and protons (~ 100 keV) precipitating from the magnetosphere with fluxes of the order of 10^{10} particles cm^{-2} sec^{-1}. Since these energetic particles come from the magnetosphere, it is natural to expect a strong conjugacy of auroral effects in the two hemispheres. This has been confirmed by actual conjugate point observations with all-sky cameras. Auroral activity shows a close correlation with all other polar disturbances such as magnetic substorms, PCA's, micropulsations, ionospheric substorms, etc. It shows also a good correlation with the *Kp* index and the sunspot number, though it has been found that the peak of the auroral activity lags the solar maximum

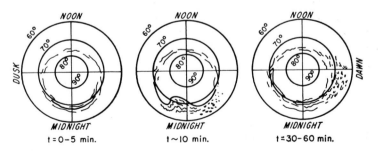

FIGURE 6.4-III The development of an auroral substorm around the auroral oval. The three diagrams show the overall distribution of auroras in the first five minutes, in about 10 minutes, and finally half an hour to an hour from the commencement of the auroral substorm

by more than one year. An interesting point is that auroras which sometimes appear inside the polar region, which is encircled by the auroral oval, show an inverse correlation with the Kp index and reach their highest intensity during magnetically quiet days. Presumably these auroras, like the quiet arcs of the auroral oval, are produced routinely by the quiet magnetosphere.

The auroral light is the sum of all the light quanta emitted by the excited, neutral and ionized, atoms and molecules, of the different atmospheric constituents. The excitation of these particles is the result of their collisions with the energetic protons and electrons which spiral down into the atmosphere along the field lines of the earth's magnetic field. The strongest spectral line of the visible aurora is the 5,577 Å line of neutral oxygen. This line is in the yellow-green region of the spectrum and gives to the polar aurora its most familiar color. Another prominent line is the doublet (6,300 Å and 6,364 Å) of neutral oxygen. This double line occurs in the red region of the spectrum where the sensitivity of the human eye is very low and therefore this radiation is nearly invisible. On some rare occasions, however, the intensity of this emission is high enough to make red auroras visible to the naked eye. These auroras are called *high altitude red arcs* (type-A auroras) and unlike most auroras which occur near 100 km, these appear around 350 km and sometimes reach altitudes up to 1,000 km.

The transition time for the 5,577 Å line is 0.74 seconds and for the 6,300–6,363 Å doublet is 110 seconds. These are immensely longer than the average time (10^{-8} sec) of typical atomic transitions and for this reason these auroral lines, like some coronal lines we discussed in Section 4.3, are called *forbidden lines*. Actually most of the auroral light comes from molecular bands mainly of neutral and ionized nitrogen, but also from oxygen. It is only the very high sensitivity of the human eye to the yellow-green light that makes the 5,577 Å line of oxygen so prominent. The same spectral sensitivity is displayed by the eyes of most living organisms, and perhaps it is an evolutionary adaptation to this trait that has made fireflies and other creatures to emit their light signals in the yellow-green region of the spectrum. Auroras with a *purplish-red lower border* (type-B auroras) owe the color of their lower border to the first positive band of neutral nitrogen molecules.

6.5 Ionospheric Disturbances

The eruption of a flare on the sun produces a series of effects in the terrestrial ionosphere. The immediate effects are due to the sudden enhancement of the ultraviolet and x-ray emission from the flare region, and their time history follows closely the development of the optical flare as seen in the red light of the Hα-line (6,563 Å). All these immediate disturbances, like the optical flare itself, appear quite suddenly and therefore are classified under the general title of *sudden ionospheric disturbances*. Since all SID's are produced by the electromagnetic radiation of the flare, they are observed only in the sunlit hemisphere.

Flare events produce also corpuscular radiation. The solar cosmic rays start arriving usually within one hour from the eruption of the flare, while the ejected plasma cloud reaches the earth in two to three days after the flare event. Both forms of solar corpuscular radiation cause many direct and indirect effects in the terrestrial ionosphere which are observed over the entire globe and last from one to three days. All in all the different ionospheric disturbances which start with the eruption of a large solar flare, can keep the ionosphere of the earth in an abnormal condition for several days.

Most of the SID's are directly related to the enhanced ionization of the D-layer, which is the result of the x-ray burst from the solar flare. During quiet days, the D-region is maintained by the Lyman-α (1,216 Å) emission from the sun and by the cosmic rays. The quiet sun emits practically no x-rays below 10 Å and therefore all the solar x-rays are stopped above the D-region. The x-ray emission from a flare, however, extends easily into the one Angstrom range which can reach the D-region and increase its electron density by at least one order of magnitude. Typical increases are from $\sim 10^3$ el/cm^3 to $\sim 10^4$ el/cm^3 at ~ 80 km and from ~ 10 el/cm^3 to $\sim 10^3$ el/cm^3 at ~ 60 km.

In the early thirties, Mögel observed occasional fadeouts in transcontinental short wave transmissions which occurred only in the sunlit hemisphere of the earth. A few years later Dellinger discovered that the fadeouts coincided with the eruption of large solar flares. Short wave radio for long distance communications uses *high frequency* (HF) waves, i.e., waves in the frequency range 3 to 30 MHz ($\lambda = 10$ to 100 m). These waves are reflected in the E- or F-layer of the ionosphere and return to the earth after passing twice through the D-layer. For stations further apart, the signals might even follow a multi-hop path which will make them cross the D-layer several times. Short wave transmissions suffer a severe loss in signal strength when an x-ray burst from the sun increases the electron density of the D-region. This phenomenon is called *short wave fadeout* (SWF) and is of great importance to telecommunications. Occasionally it might even produce a complete loss of radio signals which can last from a few minutes to nearly one hour, and is called a *radio blackout*.

Another way to measure the absorption of the D-layer is with the use of riometers (see Section 2.4). A sudden increase in the level of absorption is recorded as an abrupt decrease in the intensity of the cosmic (galactic) radio noise received. This phenomenon is called *sudden cosmic noise absorption* (SCNA) and is one of the most readily observed manifestations of an SID.

The increase in the electron density of the *D*-region during SID's improves the reflection of radio waves with wavelengths around 10 km, which propagate below the *D*-region. As a result the intensity of atmospherics, i.e. the radio noise from thunderstorm activity around the world, shows a sudden enhancement in the frequency range around 30 kHz. This phenomenon is called *sudden enhancement of atmospherics* (SEA). Longer wavelengths ($f < 20$ kHz) on the other hand, which are always reflected in the *D*-region, show little change in amplitude but display a sudden change in phase because the rapid enhancement of the *D*-layer causes an abrupt change in the height of reflection. This is called *sudden phase anomaly* (SPA) and represents one of the most sensitive means of detecting an SID.

It should be noted that in computing the highest plasma frequency f_N, up to which a signal of frequency f_m can penetrate before being reflected, we must take into consideration also the earth's magnetic field H which is represented by the cyclotron frequency f_H. The relation between f_N, f_H and f_m is,

$$f_N = f_m\left(1 + \frac{f_H}{f_m}\right)^{1/2} \qquad (6.5\text{-}1)$$

so that for $f_m \simeq 20$ kHz and $f_H \simeq 1.4$ MHz, we find $f_N \simeq 170$ kHz. This plasma frequency corresponds to $N \simeq 360$ el/cm³ which is a typical electron density in the lower *D*-region.

The increase of the electron density during an SID builds up a current system in the lower ionosphere. This current produces a transient magnetic field which is recorded by the magnetometers on the surface of the earth as a small hook (geomagnetic crochet). This phenomenon is called *solar flare effect* (SFE). The current that produces the SFE's seems to flow near the *D*-region, i.e. somewhat below the *Sq*-current, and its eddy center appears to be located closer to the subsolar point than that of the *Sq*-current.

In conclusion, the SCNA, SWF, SPA, SEA and SFE effects represent the most important manifestations of an SID and are all the results of the enhanced ionization in the *D*-region due to the x-ray burst from a solar flare. Figure 6.5-1 at the end of this section shows the nearly simultaneous occurrence of all the above mentioned effects together with the radio and optical observations of the flare. Differences in duration and relative lags in the times of the maxima occur because the ionosphere responds with a different time-constant to the different effects, and because these effects describe actually the changes at somewhat different heights of the D-region. Some of the differences are also due to the dynamic spectra (intensity vs. time at different frequencies) of the flares. The study of the different SID effects, together with satellite observations of the x-ray flux at different wavelengths, are continuously improving our understanding of the physics of the lower ionosphere and the physical processes that take place in a flare.

The effects of a flare in the E- and the F-layer are rather difficult to assess because the ionograms are very poor during a fadeout period. As a result it is very difficult to observe directly the changes of the electron density profile during an SID. Measurements, however, of the $f_0 F_2$ before and after the fadeout, show occasionally a noticeable change which is a strong indication that SID's also have a significant effect in the F-region.

Measurements of the total electron content N_T of the ionosphere, i.e., of the total number of electrons in a vertical column of unit cross-section (typically of the order of 10^{13} el/cm^2), have confirmed directly the effects of a flare in the F-region, where most of the electrons are concentrated. The N_T is usually measured by the Faraday rotation of radio signals from satellites at frequencies around 100 MHz. As a result these measurements are not affected in any serious way by absorption in the D-layer. Garriott, et al. (1967), have detected several *sudden increases of the total electron content* (SITEC) of the order of 5–10% during SID's. This implies that the solar spectrum during flares is enhanced not only in the region around 1 Å, which is responsible for the build-up of the D-layer, but also in the 10–1,000 Å range which is responsible for the ionization of the E-region and the F-region of the ionosphere.

Changes in the ionospheric layers below the maximum can be detected with still another change. Whenever absorption permits radio signals to be reflected back from the ionosphere during an SID, these signals display a minute but still detectable change in their frequency. To perform such an experiment the transmitter must be of very high stability, because for an operating frequency of about 10 MHz the frequency deviations to be measured are only of the order of 1 Hz. The phenomenon is called *sudden frequency deviation* (SFD) and is essentially a Doppler effect produced by the change in the length of the path of the reflected signals. The simplest explanation of an SFD is the sudden change of the height h of reflection. Using the well known Doppler formula $df/f = V/c$, we find that for vertical incidence in this very simple model the frequency deviation is given by the expression,

$$\frac{df}{f} = \frac{2(dh/dt)}{c} \qquad (6.5\text{-}2)$$

A similar effect is produced also by an increase in the ionization below the height of reflection, because this delays the radio signals which is equivalent to lengthening the physical path. In this case, however, the equation which describes the frequency deviation is considerably more complex than (6.5-2). In an SFD the blue shift of the frequency, i.e., lowering of the reflection height in (6.5-2), lasts for about 5minutes. Occasionally it is followed by a detectable red shift (recovery stage) which is smaller in amplitude but usually lasts considerably longer. SFD events show a good correlation with all the other effects connected with an SID, and the correlation tends to become 100% for the very large events.

Ionospheric disturbances are produced also by the corpuscular radiation from solar flare events. The first of these effects is the enhancement of radio wave absorption in the polar regions. This is produced by energetic protons from the sun (solar cosmic rays) which are deflected toward the polar caps by the earth's magnetic field. The phenomenon is usually observed with riometers and is called *polar cap absorption* (PCA). In the language of energetic particles, PCA's are called proton events. The absorption occasionally is so strong that no trace can be seen in the ionograms of the ionospheric sounders. Such a no-echo case of total absorption in the polar regions is called a *polar blackout*. The magnitude of the absorption is sometimes measured in terms of the lowest frequency which produces an echo in the ionograms. This frequency is called f_{min}. PCA's usually start half an hour to a few hours after the x-ray burst. They reach maximum intensity only a few hours after their onset and then they remain at a high, slowly decreasing level for one or more days.

The arrival of the plasma cloud from the flare event, one or two days after the eruption of the flare, signals the beginning of a geomagneti cstorm. During the storm we usually have several polar substorms which are accompanied by ionospheric disturbances in the polar regions. The most noticeable of these disturbances is the enhancement of the absorption in the auroral region. This effect is called *auroral zone absorption* (AZA) or *auroral blackout* and, like the polar substorms, lasts from 1 to 3 hours. The AZA is caused by energetic electrons (20–100 keV) which accompany the less energetic electrons (~ 10 keV) which produce the auroras. As the energetic electrons descend into the atmosphere, they generate x-rays through bremsstrahlung, which in turn enhance the ionization of the ionosphere. Auroral absorption events occur often simultaneously at two conjugate points in the northern and southern auroral ovals.

During the days of a magnetic storm, the entire ionosphere is in a disturbed condition and therefore parallel to the magnetic storm we experience an *ionospheric storm*. This can be seen from the good correlation which the intensity of the different ionospheric disturbances, such as spread-F, sporadic-E and ionospheric irregularities, shows with the 3-hour planetary index of magnetic activity (Kp-index). It has also been suggested that the *internal gravity waves* which are generated in the atmosphere during polar substorms produce the large travelling ionospheric disturbances which we discussed in Section 2.6. According to this theory, the travelling ionospheric disturbances are produced at two magnetically conjugate locations in the aurora region and then travel toward the equator with speeds of about 300–400 m/sec.

During an ionospheric storm we can observe substantial changes in the critical frequency f_0F_2, the total content N_T, the height of the maximum h_{max}, and in general of the entire electron density profile of the ionosphere. A major cause of all these variations is the increase in the temperature of the atmosphere and the ionosphere which takes place during the storm. The principle heat source for this increase has not been established yet, but some of the most probable candidates are:

A   The impinging auroral particles and the energetic particles injected into the zones of trapped radiation.

B   Heat conduction from the hot plasma cloud which engulfs the earth.

C   The dissipation, as heat, of the different electric currents which are generated during the storm.

D   The hydromagnetic waves which are produced by the geomagnetic storm and dissipate their energy as they propagate through the ionosphere.

An increase in the temperature changes substantially the electron density profile of the ionosphere, not only because of the expansion of the ionospheric plasma (a higher temperature means a larger scale height), but also because of the change in the recombination rate at the different altitudes. The recombination rate is represented by the attachment coefficient β (see Section 2.5) which depends on the local density and energetic excitation of the neutral particles. Both of these properties are temperature sensitive but their relative effects vary with height and with the value of the initial temperature. As a result one can expect to observe seasonal and latitudinal effects in the changes of the ionospheric profile during geomagnetic storms.

The electron density of the upper ionosphere can also increase by the dumping of ionization from the protonosphere into the ionosphere. The magnitude of this effect depends primarily on the intensity of the magnetic disturbance, but is also depends on the geomagnetic latitude of the area and on the local time, reaching a maximum near sunset on the first day of the storm (Papagiannis, Mendillo, and Klobuchar, 1971).

It is well known that a substantial amount of the total ionization is stored in the plasmasphere along the tubes of force of the earth's magnetic field. Whistler observations (Carpenter, 1970) show that during magnetic storms these tubes are seriously depleted and the plasmapause moves considerably closer to the earth (from $\sim 5.5 R_0$ to $\sim 3.5 R_0$).

Whistlers are low frequency radio noise signals from lightning discharges which in the 1–10 kHz range can travel between conjugate points along the lines of force of the earth's magnetic field. The different frequency components of these signals travel at different velocities in the plasma of the exosphere. This dispersion (time delay with frequency) allows us to determine the electron density near the crossing of the equatorial plane by the field lines along which these signals were travelling.

Changes in the electron density can be produced also by horizontal drifts of ionization. Large temperature gradients at ionospheric heights produce *neutral winds*, i.e., horizontal motions of the neutral particles of the atmosphere, which through collisions drag along the plasma particles of the ionosphere. These winds can induce also vertical motions of the ionization, especially at high latitudes, because the charged particles prefer to move along the lines of force of the earth's magnetic field. Motions of ionization can be produced also by $E \times B$ forces, where B is the earth's magnetic field and E one of the electric fields that generate the different electric currents in the ionosphere during magnetic storms.

In conclusion, it is fair to say that the world-wide behavior of the ionosphere during ionospheric storms is one of the most complicated subjects in the field of solar-terrestrial relations. The overall picture is still incomplete and so is our understanding of the different effects so far observed. A well coordinated collection of data from ground stations around the world as well as from satellites and space probes, together with further theoretical studies, will undoubtably improve our understanding of this intriguing field in the years to come.

The very good international cooperation which now exists in this field (best exemplified by the establishment in 1967 of an International Commission on Solar-Terrestrial Physics) will undoubtably be of great help in coordinating our efforts.

An important new element which has now come into the picture is that we have started finally to combine seemingly unrelated observations (radio, x-ray, solar, geophysical, etc.) into a single framework which engulfs the entire physical problem. This integrated approach is probably the most important new development in ionospheric physics and solar physics and has practically revolutionized the field of solar terrestrial relations. In the light of our discussion at the end of Section 4.7 about "typical events",

we have tried to summarize in Figure 6.5-I the chronological development
of the different effects which are observed on the earth following the eruption of a rather large flare on the sun.

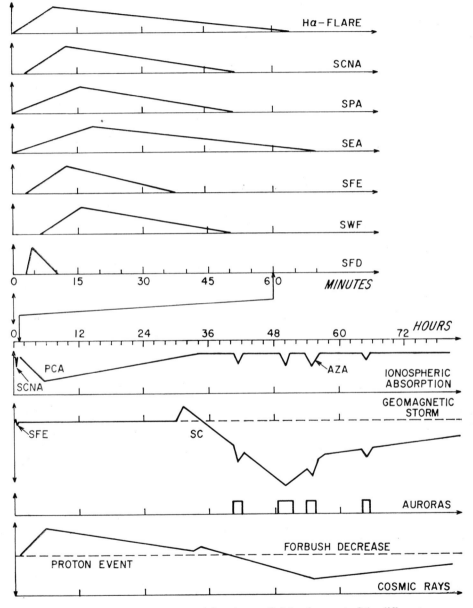

FIGURE 6.5-I   A diagram summarizing the parallel development of the different
effects in the area of solar-terrestrial relations

6.6 Bibliography

A. Books for Further Studies

1. *Polar and Magnetospheric Substorms*, Syun-Ichi Akasofu, D. Reidel Publ. Co., Dordrecht, Holland, 1968.
2. *Solar Plasma Geomagnetism and Aurora*, S. Chapman, Gordon and Breach, New York, N. Y., 1964.
3. *Solar Radio Astronomy*, M. R. Kundu, John Wiley & Sons, New York, N. Y., 1965.
4. *Astrophysics and Space Science*, Allen H. McMahon, Prentice Hall, Englewood Cliffs, N. J., 1965.
5. *Solar Flares*, H. J. Smith and E. V. P. Smith, The MacMillan Co., New York, N. Y., 1963.
6. *Radio Astronomical and Satellite Studies of the Atmosphere*, ed. by J. Aarons, North-Holland Publ. Co., Amsterdam, Holland, 1963.
7. *High Latitude Particles and the Ionosphere*, ed. by B. Maehlum, Academic Press, New York, N. Y., 1965.
8. *Physics of Geomagnetic Phenomena*, ed. by S. Matsushita and W. H. Campbell, Academic Press, New York, 1967.
9. *Space Physics*, ed. by D. P. LeGalley, John Wiley & Sons, New York, N. Y., 1964.
10. *Research in Geophysics*, ed. by H. Odishaw, the MIT Press, Cambridge, Mass., 1964.
11. *Introduction to Space Science*, ed. by W. Hess and G. D. Mead, Gordon and Breach, New York, N. Y., 1968.
12. *Handbook of Geophysics and Space environments*, ed. by S. L. Valley, AFCRL, Bedford, Mass., 1965.
13. *Physics of the Earth's Upper Atmosphere*, ed. by C. O. Hines, I. Paghis, T. R. Hartz and J. A. Fejer, Prentice-Hall, Englewood Cliffs, N. J., 1965.
14. *Introduction to Solar Terrestrial Relations*, ed. by J. Ortner and H. Maseland, D. Reidel Publ. Co., Dordrecht, Holland, 1965.
15. *Solar Terrestrial Physics*, ed. by J. W. King and W. S. Newman, Academic Press, New York, N. Y., 1967.
16. *Physics of the Magnetosphere*, ed. by R. L. Carovillano, J. F. McClay and H. R. Radoski, D. Reidel Publ. Co., Dordrecht, Holland, 1968.
17. *Earth's Particles and Fields*, ed. by B. M. McCormac, Reinhold Book Corp., New York, N. Y., 1968.
18. *Magnetospheric Physics*, ed. by D. J. Williams and G. D. Mead, Amer. Geophys. Union, Washington, D. C., 1969.
19. *The Solar Spectrum*, ed. by C. de Jager, D. Reidel Publ. Co., Dordrecht, Holland, 1965.

B. Articles in Scientific Journals

Cahill, L. J., The inflation of the inner magnetosphere during a magnetic storm, *J. Geophys. Res.*, **71**, 4505, 1966.

Carpenter, D. L., Whistler evidence of the dynamical behavior of the dusk side bulge in the plasmasphere, *J. Geophys. Res.*, **75**, 3837, 1970.

Dessler, A. J., and E. N. Parker, Hydromagnetic theory of geomagnetic storms, *J. Geophys. Res.*, **64**, 2239, 1959.

Filz, R. C., *et al.*, Corpuscular radiation; A revision of Chapter 17, *Handbook of Geophysics and Space Environments*, AFCRL-68-0666, Dec., 1968.

Föppl, H. *et al.*, Artificial strondium and barium clouds in the upper atmosphere, *Planet. Space Sci.*, **15**, 357, 1967.

Frank, L. A., On the extraterrestrial ring current during geomagnetic storms, *J. Geophys. Res.*, **72**, 3753, 1967.

Frank, L. A., Direct detection of assymmetric increases of extraterrestrial ring current proton intensities in the outer radiation zone, *J. Geophys. Res.*, **75**, 1263, 1970.

Forbush, S. E., Variations with a period of two solar cycles in the cosmic-ray anisotropy and the superposed variations correlated with magnetic activity, *J. Geophys. Res.*, **74**, 3451, 1969.

Garriot, O. K., A. V. DaRosa, M. J. Davis and O. G. Villard, Solar flare effects in the ionosphere, *J. Geophys. Res.*, **72**, 6099, 1967.

Obayashi, T., Propagation of solar corpuscles and interplanetary fields, *J. Geophys. Res.*, **67**, 1717, 1962.

Papagiannis, M. D., M. Mendillo, and J. A. Klobuchar, Simultaneous increases of the ionospheric total electron content and the geomagnetic field in the sunset sector, *Planet. Space Sci.*, **19**, 503, 1971.

Parker, E. N., Theory of streams of cosmic rays and the diurnal variation, *Plan. Space Sci.*, **12**, 735, 1964.

Vasyliunas, V. M., A survey of low energy electrons in the evening sector of the magneto-sphere with OGO-1 and OGO-3, *J. Geophys. Res.*, **73**, 2834, 1968.

CHAPTER 7

SOLAR AND PLANETARY SPACE ASTRONOMY

7.1 The Domain and the Scope of Space Astronomy

For many millenia, people mystified by the splendor of the sun and the night sky have tried to understand the universe that surrounds us. The inquisitive observer of the early days used only his eyesight, and therefore his observations were confined in the visible region (4,000–7,000 Å) of the electromagnetic spectrum. These early astronomical observations were made possible only thanks to the fact that the spectral sensitivity of the human eye nearly matches the singular, and almost equally narrow window (3,000–10,000 Å) of the terrestrial atmosphere. Optical observations, especially with the help of telescopes and photographic techniques, have allowed us to accumulate an impressive amount of knowledge about the astronomical universe. In the last 30 to 40 years, however, we began to realize that we can learn much more about the cosmos by extending our astronomical observations to the other regions of the spectrum.

One can infer the rich crop which awaits the explorers of new spectral regions from the tremendous gains of radio astronomy. This new window of the terrestrial atmosphere was discovered in 1932 by Karl Jansky who realized that his rather primitive radio antenna at the Bell Telephone Laboratories was receiving radio noise of extraterrestrial origin. This noise turned out to be radio emission from the center of our galaxy (see Section 8.1). In the few decades since this accidental discovery, radio astronomy has grown extremely fast and now competes on equal terms with optical astronomy. The wealth of new knowledge which we have gained through radio astronomy includes the discovery of radio emission from Jupiter, which has revealed the strong magnetic field of this planet; the detection of radio bursts from the sun, which have advanced decisively our knowledge of flare events; and the mapping of the Milky Way at the 21-cm line of atomic hydrogen, which has given us the first clear indication of the spiral structure of our galaxy. The contributions of radio astronomy have been even more spectacular in the last decade. Pulsars, quasars, interstellar molecules, and the 3°K background radiation (see Section 8.1) are some of the more recent discoveries of radio astronomy. It is important to add here that quasars became the truly exciting objects we know today only when radio and optical observations were combined together.

The accomplishments of radio astronomy and the realization of the great value of parallel observations at many different wavelengths steered the attention of many scientists to the still unexplored regions of the spectrum. These were zealously kept hidden by the terrestrial atmosphere, but the advent of the space age allowed us finally to conquer this formidable barrier. (The interested reader will be able to find in Appendix II a brief recount of the development of the space age.) The ability which we have finally gained to conduct observations above the obstructing layers of the terrestrial atmosphere has created an entirely new division of astronomy. This new division is called *space astronomy* and is very different from the classical *ground-based astronomy*.

Space astronomy is subdivided into several branches according to the spectral domain of the observations. The problems, the instruments, and the specific objectives in each spectral region, however, are quite distinct and as a result these subdivisions have evolved almost as independent fields. Thus space astronomy consists now of several "astronomies" namely: *γ-ray astronomy, x-ray astronomy, ultraviolet astronomy, infrared astronomy*, and *space radio astronomy*. It should be noted here that infrared astronomy has also a foot on the ground, because some very interesting observations can, and are made now from the ground through the different infrared windows of the terrestrial atmosphere. By the same token, however, *optical astronomy* has a lot to gain through space-borne telescopes and therefore in the subdivisions of space astronomy we must also include the space branch of optical astronomy.

Gamma rays are stopped in the denser part of the atmosphere through Compton scattering and pair production. The x-rays are stopped at the higher layers of the atmosphere where they ionize the different atmospheric constituents. The same is also true for the ultraviolet radiation at $\lambda < 900 \text{Å}$. The UV rays in the range between 900 and 2,000 Å are strongly absorbed by the molecular oxygen of the atmosphere, producing its photodissociation. Finally the UV rays in the 2,000 to 3,000 Å range are absorbed by the layer of ozone (O_3) which, as a minor constituent, envelopes the earth at heights between 20 and 50 km.

The infrared radiation, except for a few minor windows in the range between 1 and 25 microns (see Section 8.5), is absorbed by the rotational and vibrational bands of different atmospheric constituents such as carbon dioxide (CO_2) and water vapor (H_2O). Finally, the long radio waves beyond the far end of the radio window, i.e., radio waves with λ greater than about

13   Papagiannis

30 meters, are reflected back by the terrestrial ionosphere. The wavelength λ_c (in meters) at the cutoff point is given by the expression,

$$\lambda_c = \frac{10^5}{3\sqrt{N_m}} \qquad (7.1\text{-}1)$$

where N_m is the maximum electron density of the ionosphere in el/cm³. Figure 7.1-I is a diagram of the electromagnetic spectrum showing the optical and the radio windows of the atmosphere and the vast regions of the spectrum which, for the reasons discussed above, are not accessible to the ground-based observer.

Figure 7.1-II shows the minimal heights at which observations in these inaccessible regions of the spectrum become possible. Of course the quality of all space observations continues to improve as we move to higher altitudes. This holds true even for optical observations because, though

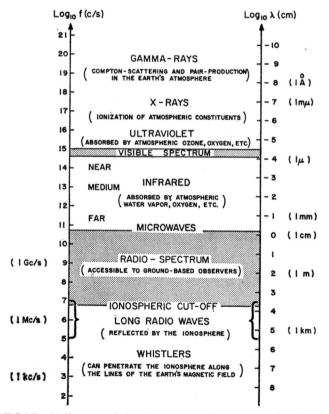

FIGURE 7.1-I A diagram of the electromagnetic spectrum showing the only two spectral regions (the optical window and the radio window) which are available for ground-based astronomical observations

the atmosphere is practically 100% transparent in the visible region, the *turbidity* (turbulence) of the atmosphere degrades strongly the resolving power of groundbased telescopes. It should be added here that the terrestrial atmosphere constitutes also an opaque barrier for the cosmic rays which are of great astrophysical interest.

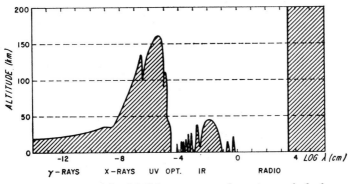

FIGURE 7.1-II   The minimal heights necessary for astronomical observations in the different regions of the electromagnetic spectrum

In the following sections of this chapter we will discuss the solar and planetary aspects of space astronomy, and in the next chapter, after a brief introduction to galactic astronomy, we will discuss the observations, the theory, and the prospects of galactic space astronomy. In both cases we will examine separately the different spectral regions following a sequence from the shorter to the longer wavelengths.

There are, however, some common and unifying objectives of space astronomy which we can mention in advance because they remain the same throughout the entire spectrum. The most important of these are:

a   The detection of new sources and the identification with their counterparts in other regions of the spectrum.

b   The measurement of the emission spectra, polarization, and spectral lines over as wide a spectral range as possible.

c   The study of temporal variations in the emission characteristics of different astronomical objects.

d   The study of the diffuse background radiation in all spectral regions.

The combination of a, b, and c will lead to comprehensive physical models of the many celestial objects under investigation, and d will help us answer certain basic cosmological problems. The almost unlimited spectral horizons of space astronomy promise a very bright future for astronomy in general in the years to come.

13*

7.2 Solar X-Ray Astronomy

The first concrete evidence of x-ray emission from the sun was obtained in 1949 by Herbert Friedman and his co-workers of the Naval Research Laboratory with two Geiger counters on a captured German V-2 rocket. Solar x-ray astronomy extends from around 100 Å, where the extreme ultraviolet ends, to about 0.02 Å where the γ-ray spectrum begins. The x-ray spectrum is often divided into the *soft* and *hard* x-ray regions, the dividing line commonly placed around 1 Å. It should be noted, however, that all these spectral divisions are rather arbitrary and vary considerably from one author to another depending on his theoretical interests and his observing technique.

At the longer wavelength region of the x-ray spectrum we are still in the domain of optics where photons are characterized by their wavelength in Angstroms. At the more energetic segment of the x-ray spectrum, however, we are already in the domain of nuclear physics where photons are usually characterized by their energy in kiloelectronvolts. The relation between the two units is simply,

$$\lambda(\text{Å}) \times E(\text{keV}) \simeq 12.4 \qquad (7.2\text{-}1)$$

The solar corona at a temperature of approximately $10^{6\circ}$K is a natural emitter of x-rays which can account for the formation of the E-layer of the ionosphere. From Wien's displacement law ($T \times \lambda_{\text{max}} \simeq 0.3$) we find that the maximum x-ray emission from the quiet sun must be around 20–30 Å. The result is essentially correct, though we are not dealing here with a black body since in the x-ray region the solar corona is optically thin (nearly transparent). The x-ray emission of the solar corona is primarily due to the integrated flux of many x-ray emission lines. This line flux exceeds by at least a factor of ten the continuous thermal x-ray emission of the solar corona.

The x-ray emission level of the solar corona contains a slowly varying component because it varies with the 11-year cycle of the sun. It also varies with the 27-day rotation of the sun and the active centers on it. Active regions in the corona, with temperatures of 2–$4 \times 10^{6\circ}$K and densities several times higher than the densities of the quiet corona, emit a considerably enhanced x-ray flux especially in the 5 to 10 Å range where the x-ray emission of the quiet corona is practically non-existent (Figure 7.2-I). These active regions can last up to 2–3 solar rotations and at solar maximum there might be several of them on the sun at the same time.

Superimposed on the slowly varying component there is a fast transient component of solar x-ray emission which is associated with flare events. These x-ray bursts, which are usually well correlated with radio bursts, were first observed in 1956 and by now their morphology has been studied to a considerable extent. In Section 4.6 we discussed the characteristics (production mechanisms, duration, etc.) of the two known types (Class I and Class II) of solar x-ray bursts, so there is no need to repeat this description here. Solar x-ray astronomy is of great interest not only to solar astronomers but also to space physicists because, as we have seen in Section 6.5, solar x-ray bursts produce many important effects (radio blackouts, radiation hazards for astronauts, etc.) in our immediate environment.

A model for the source of solar x-ray bursts was developed by Strauss and Papagiannis (1971). In this model, which is shown in Figure 7.2-II, the field lines of a bipolar sunspot group form closed loops only up to a certain height. Above this height, the field lines are drawn out into the interplanetary space by the solar wind where they form a coronal streamer. Due to certain instabilities, the magnetic energy, which is stored in the stretched field lines of the streamer, is released and is used to accelerate charged particles to high energies. As seen from Figure 7.2-II, some of these high energy particles move outwards into the coronal streamer, where they produce a type III solar radio burst, and some move inwards toward the lower corona where they emit an impulsive (Class II) x-ray burst as they are being injected at the top of the loop.

The impulsive x-ray burst is due to non-thermal bremsstrahlung which is emitted as the collimated beam of these high energy particles is stopped by the denser plasma of the loop. The stream of these energetic particles lasts for approximately a few minutes and its peak corresponds to the

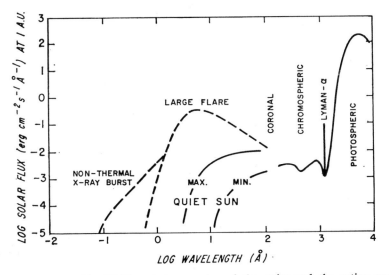

FIGURE 7.2-I   The XUV emission spectra of the quiet and the active sun

non-thermal Class II x-ray bursts which are usually seen in the 0.1–0.5 Å range (Figure 4.6-III and Figure 7.2-II). The thermalization of the high energy beam deposits a large amount of energy at the top of the loop. This heat source can be approximated by the expression,

$$Q(s, t) = Q(0, t) \exp\left(-\frac{s^2}{2\sigma^2}\right) \text{erg cm}^{-3} \text{sec}^{-1} \qquad (7.2\text{-}2)$$

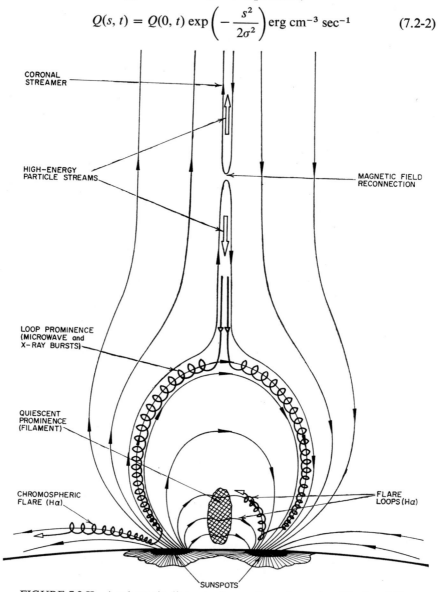

FIGURE 7.2-II A schematic diagram of an active solar region. The instability develops in the field lines from a bipolar sunspot group which are stretched by the solar wind into the interplanetary space (Strauss and Papagiannis, 1971)

where t is the time, s is the arc length from the top to either side of the loop, and σ is the stopping distance of the energetic particles in the plasma of the loop. Heat is conducted symmetrically along both sides of the loop, which is assumed to have a cross-section $A(s)$.

Very little energy is conducted to the outside of the loop because the thermal conductivity of a hot plasma is much higher along the magnetic field lines than across them. Through this process the plasma of the loop becomes heated to temperatures of about $3\text{--}5 \times 10^7 °\text{K}$ and starts emitting x-rays through thermal Bremsstrahlung. This is the non-impulsive (Class I), thermal x-ray burst which peaks around $1\text{--}5$ Å and lasts from less than 10 minutes to more than one hour. The duration of the thermal x-ray burst represents the period over which the plasma of the loop remains hot enough to produce a substantial x-ray emission and can be estimated by computing the heat losses from the loop.

The energy lost by this hot plasma is primarily due to radiative cooling and therefore the loss term is given by the expression,

$$R = N^2 f(T) \text{ erg cm}^{-3} \text{ sec}^{-1} \tag{7.2-3}$$

where N is the electron density ($\sim 3 \times 10^{11}$ el/cm^3) in the loop and $f(T)$ is a known function of the temperature, approximately proportional to $T^{1/2}$. Since, as we mentioned, the energy is conducted primarily along the length of the loop, this becomes essentially a one dimensional problem and the heat conduction equation takes the form,

$$3Nk \frac{dT(s, t)}{dt} = \frac{1}{A(s)} \frac{\partial}{\partial s} \left[A(s) \, K(T) \frac{\partial T(s, t)}{\partial s} \right] + Q(s, t) - R(s, t) \tag{7.2-4}$$

where $K(T)$ is the thermal conductivity of a fully ionized plasma (5.2-14), which is proportional to $T^{5/2}$. Assuming a constant cross-section A, and introducing in (7.2-4) the expressions for Q and R from (7.2-2) and (7.2-3) we obtain the relation,

$$\frac{\partial T(s, t)}{\partial t} = \frac{1}{3Nk} \frac{\partial}{\partial s} \left[K(T) \frac{\partial T(s, t)}{\partial s} \right] + \frac{Q(0, t)}{3Nk} e^{-s^2/2\sigma^2} - \frac{Nf(T)}{3k} \tag{7.2-5}$$

which is independent of the cross-section A. Thus A is a parameter which does not affect the time evolution of the x-ray burst but can be used as a scaling factor for the intensity of the burst. Equation (7.2-5) is a non-linear partial differential equation which can be solved numerically on a computer with the method of finite differences to give the temperature T

as a function of s and t. Knowing T along the entire loop, we can use the volume free-free emission coefficient,

$$j_\nu(T) = 5.44 \times 10^{-39} g N^2 T^{-1/2} e^{-(h\nu/kT)} \text{ erg cm}^{-3} \text{ sec}^{-1} \text{ Hz}^{-1} \text{ sr}^{-1} \qquad (7.2\text{-}6)$$

(g is the so-called Gaunt factor usually of the order of unity) to find the total x-ray emission in a given energy band at any instant of time. The theoretical results of (7.2-5) were found to be in good agreement with actual satellite data of solar x-ray bursts and can reproduce to a very acceptable degree the evolution of thermal (Class I) x-ray bursts at different energy bands of the x-ray spectrum.

The most commonly used x-ray detectors are of the gas-filled type, such as Geiger counters and proportional counters. Scintillation counters have also been used at the energetic (~ 100 keV) end of the spectrum. Most of these detectors are used in conjunction with filters which provide some spectral sensitivity but still give only the integrated flux over relatively broad wavelength bands (e.g., 2–8 Å). In order to study the emission lines of the x-ray spectrum, we need much higher spectral resolution. At wavelengths longer than about 25 Å this is achieved with diffraction gratings at grazing angles of incidence, while below 25 Å spectroscopic observations are made with Bragg crystal spectrometers. As seen from Figure 7.2-III, for a given angle of incidence θ and a given spacing d of the atomic layers of the crystal, only wavelengths that satisfy the relation,

$$n\lambda = 2d \sin\theta \qquad (7.2\text{-}7)$$

will be reflected and focussed on the detector. The spacing d is typically of the order of 10 Å and the instruments are normally used with $n = 1$. By varying the angle θ we can obtain a detailed scanning of the spectrum. Resolutions of the order of $\lambda/\Delta\lambda = 1,000$ or better, have been achieved in the range around 10 Å.

Theoretical computations show that the solar spectrum might also contain some lines in the near γ-ray region such as the 0.51 MeV line resulting from the annihilation of positrons with electrons, and the 2.23 MeV line

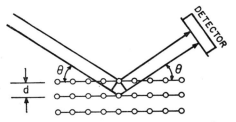

FIGURE 7.2-III   A schematic diagram of the Bragg crystal spectrometer which is used for line studies in the x-ray region

produced by the capturing of neutrons by protons to form deuterium. The positrons and the neutrons needed in these reactions are produced through the inelastic collisions and spallation reactions of flare accelerated protons (Dolan and Fazio, 1965). These γ-ray lines from the sun have not been detected yet, but their eventual detection and measurement will help us to determine the energy spectrum of the flare protons and the nuclear reactions that might possibly take place in flare events.

In solar astronomy it is of great importance to measure the change in the intensity of the emitted radiation from the center to the limb of the solar disc, and to be able to study in detail the morphology of the different active regions. Such studies naturally require high angular resolution. X-ray images of the sun were first obtained by the NRL group in the early sixties using a pinhole camera with an x-ray sensitive emulsion. Visible and infrared radiation were kept out by placing a thin, aluminum covered, plastic foil over the pinhole. More recently, however, the development of x-ray telescopes has allowed us to obtain x-ray pictures of the sun of much higher quality (Vaiana *et al.*, 1968).

X-rays can be specularly reflected from mirror surfaces at grazing incidence angles and therefore the inner surface of a slightly tapered cylinder can be used as a focussing device (telescope) for x-rays. To avoid aberration effects x-ray telescopes use in combination a paraboloid and a hyperboloid (Figure 7.2-IV) and the x-rays are reflected from both of these surfaces before they are focussed on a detector or a photographic plate.

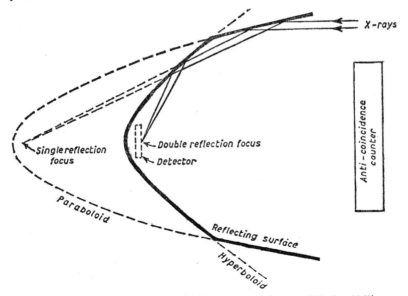

FIGURE 7.2-IV   A grazing incidence x-ray telescope (Weeks, 1969)

The image of the sun at different x-ray bands can be obtained with the use of filters in conjunction with x-ray telescopes. These images are called *x-ray spectroheliograms* and are being obtained now on a routine basis by the different OSO satellites. Significant in this respect has been the contribution of several European groups. A team, e. g., from University College and the University of Leicester, under Profs. Boyd and Stewardson, obtained with OSO-V x-ray images of the sun at seven 1–2 Å bands in the 8–18 Å range with a resolution of 1.6 arc minutes, while Prof. Brini's group of the University of Bologna operated a 20–200 keV spectroheliograph on OSO-VI. The combination of large x-ray telescopes with Bragg crystal spectrometers will make possible in the future x-ray spectroheliograms of much higher angular and spectral resolution.

7.3   Ultraviolet, Optical, and Infrared Solar Space Astronomy

The ultraviolet (UV) spectrum extends from about 100 Å to approximately 4,000 Å. The region between roughly 100 and 1,000 Å is often called the *extreme ultraviolet* (EUV), while the region between 3,000 and 4,000, which is accessible to ground-based observers, is sometimes referred to as the *near ultraviolet*. The first ultraviolet spectrum of the sun beyond the 3,000 Å cut-off of the terrestrial atmosphere was obtained in 1946 by Richard Tousey and his co-worker at the Naval Research Laboratory using a German V-2 rocket. The progress in solar UV astronomy since these early pioneering days has been truly impressive.

The UV is probably the most interesting region of the solar spectrum for many reasons. First of all it contains a very large number of absorption and emission lines from all the elements (in their different stages of excitation and ionization) that are found in the atmosphere of the sun. It is also of great value because it is produced in a region of the solar atmosphere (chromosphere—transition zone—lower corona) which still holds many puzzling questions. Finally, it contains a wealth of information on solar activity and its different lines can act as thermometers at different depths in the atmosphere of the sun.

The intensity of a given line is essentially proportional to the total number of atoms which can emit this line. These are the atoms that happen to be in a particular state of excitation and ionization. Let N_i be the number density of atoms in a given excited state, and N the total number of atoms in the state of ionization in which this excited state belongs (all the atoms of a given element that have the same number of electrons arranged in any possible way). The ratio of N_i to N is given by the *Boltzmann equation*,

$$\frac{N_i}{N} = \frac{g_i}{U(T)} \exp\left(-\varepsilon_i/kT\right) \tag{7.3-1}$$

where ε_i is the excitation energy of this state, k the Boltzmann constant (1.38×10^{-16} erg/degree), T the absolute temperature, g_i the *statistical weight* of this excitation state, and $U(T)$ the *partition function* for all atoms in this state of ionization.

In a rather mundane analogy it is like asking how many of the people living in a hotel are staying on each floor, assuming that the hotel has no elevators. Naturally most people would tend to occupy the lowest floors (smaller ε_i), except when it gets very hot (high T) and they want to have more fresh air. Also the final distribution will depend on the number of rooms available on each floor (g_i) compared to the total number of rooms (U) that the hotel has. The quantities g_i, $U(T)$, and ε_i are generally known and tabulated so that the appearance of a certain line reveals the temperature T in the region where the line is emitted. Furthermore if N is also known the intensity of the line reveals the total volume where the temperature is at this level.

To find, for a given element, the ratios of atoms in the different states of ionization we must use the *Saha equation*. If, e.g., NII is the number of singly ionized atoms and NI the number of neutral atoms of this element, then Saha's equation says that,

$$\frac{N\text{II}}{N\text{I}} N_e = \frac{(2\pi mkT)^{3/2}}{h^3} \frac{2U\text{II}(T)}{U\text{I}(T)} \exp\left(-E_{\text{I,II}}/kT\right) \tag{7.3-2}$$

where $E_{\text{I,II}}$ is the ionization potential (the energy needed to singly ionize this element), UI and UII are the partition functions for the neutral and ionized state; and N_e is the ambient electron density. The Saha equation can be applied also for any two consecutive states of ionization, e.g., X and XI, by simply using in (7.3-2) the corresponding partition functions and ionization potential. Usually the energy difference between the different states of ionization is such that practically all the atoms of a given element are found in one, or at most in two consecutive states of ionization.

As seen from (7.3-1) and (7.3-2) the intensity of a line depends strongly on the temperature and since in the solar chromosphere and lower corona the temperature varies rapidly with height, the altitude range from which a specific line is emitted is quite narrow and well-defined. Thus a spectro-heliogram gives the conditions at a specific height of the solar atmosphere and a sequence of such monochromatic images of the sun at the wavelengths of different lines can provide a complete height profile of the solar atmosphere over the entire disc of the sun.

Starting from the Lyman continuum around 900 Å, which is formed at about 10,000°K, and proceeding with ultraviolet emission lines such as N III ($\sim$100,000°K), O VI ($\sim$300,000°K), Si VI ($\sim$500,000°K), Si VIII ($\sim$900,000°K), and Mg X ($\sim$1,400,000°K), we can obtain a well-stratified picture of the chromosphere, the steep transition zone, and the lower corona of the sun. These observations are of unique value in our effort to understand the physics of this intricate region of the solar atmosphere.

The sixth *Orbiting Solar Observatory* (OSO-VI), a satellite from this series is shown in Figure 7.3-I, carried on board a sophisticated UV spectroheliograph which could obtain monochromatic pictures of the sun at any wavelength in the range 300 to 1400 Å. This instrument, which was constructed and operated by Professor Goldberg's group of Harvard University, could scan in a few minutes the entire disc of the sun with an angular resolution of 35 arc seconds. The data, in the form of a computer printout, reached the scientists at the Harvard Observatory in almost *real time* (in space language the time of the actual observation). Figure 7.3-II is a typical example of these data showing in this case the intensity pattern of the solar disc in the 625 Å line of Mg X.

On a command from the ground the spectroheliograph could switch to a different wavelength, or would make a higher resolution picture of a small

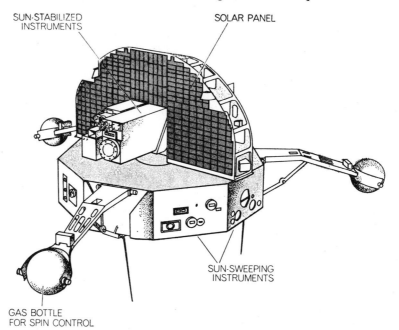

FIGURE 7.3-I   A drawing of one of the Orbiting Solar Observations (OSO) satellites (Goldberg, 1969)

area of the solar disc for a more detailed study of an active region. It also could be ordered to obtain the solar spectrum in the 1400 to 300 Å range. Figure 7.3-III shows such a spectral scan from the center of the solar disc. One can see several important lines such as the Lyman-α at 1,216 Å and the He II line at 304 Å. Also clearly seen is the Lyman continuum, which starts at 912 Å and decreases in intensity with decreasing wavelength.

The spectrum of the sun in the UV is known with a resolution of better than 0.1 Å. Many thousands of lines have been discovered from all the relatively more abundant elements (H, He, O, N, C, Si, Mg, Ne, Fe, etc.) of the solar atmosphere in their many possible states of excitation and ionization. The intensity of some of these lines, such as the Si XII and the Fe XVI, changes by a factor of 10 or even 100 in regions of high solar activity. As a result a sequence of high speed and high resolution UV spectroheliograms of an active region can provide extremely valuable data on the development of a flare event.

FIGURE 7.3-II The intensity pattern of the solar disc in the 625 Å line of Mg X, drawn from a computer print-out of Harvard's OSO-VI data (Papagiannis, 1969)

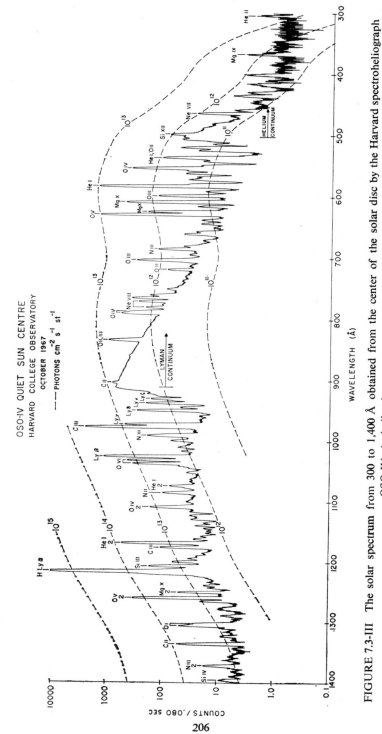

FIGURE 7.3-III The solar spectrum from 300 to 1,400 Å obtained from the center of the solar disc by the Harvard spectroheliograph on OSO-IV. A similar instrument was carried also on OSO-VI

Most of the emission lines from the chromosphere and the lower corona occur at wavelengths shorter than 1,500 Å where the emission from the photosphere is practically negligible. As a result these emission lines can be observed without difficulty in front of the solar disc. This cannot be done with most of the chromospheric and coronal lines in the optical domain where these lines are swamped by the intense photospheric emission and therefore they can be seen only beyond the limb of the sun.

Ultraviolet observations will also provide valuable data on the density and temperature profiles of the solar atmosphere through the study of the limb brightening effects in the different UV emission lines, and by studying the center to limb variation in the ratio of two parts of the Lyman continuum. We hope also to obtain more accurate values for the abundances of the different elements in the solar corona, and through high resolution spectroheliograms to study the existence of fine filamentary structures in the solar corona. Finally high spectral resolution studies of the Zeeman splitting of coronal lines will allow us to measure the magnetic fields of the solar corona and relate them to corresponding field configurations on the surface of the sun.

In the optical window (3,000–9,000 Å) the terrestrial atmosphere is essentially transparent but its turbulence causes a serious problem (*seeing*) for ground-based observations. The rapid appearance and disappearance of small scale atmospheric inhomogeneities changes continuously the index of refraction along the path of the optical rays and produces the twinkling (scintillation) of the stars and a random motion (dancing) of the telescopic images around their actual position. This last effect degrades the sharpness of the image and limits the angular resolution of all telescopes to about 0.5 seconds of arc. This is way below the theoretical limit of our large telescopes and can be easily achieved by a ten-inch telescope above the turbulent layers of the terrestrial atmosphere. The Rayleigh scattering of the sunlight by the molecules of the atmosphere is one more problem for ground-based optical observers. This is responsible for the bright blue color of the sky which makes impossible any stellar or galactic optical observations during the day and strongly degrades the contrast of solar observations.

In 1957, a Princeton University group under the direction of Martin Schwartzschild launched a balloon carrying a 12-inch reflector telescope (Stratoscope-I) to photograph the sun above the turbulent layers of the stratosphere. Stratoscope I reached altitudes close to 80,000 ft ($\sim$24 km) and in its five flights, especially in its last one in 1959, it obtained excellent pictures of the solar granulation as well as of sunspots which show clearly the filamentary structure of the penumbra of the spots.

Balloon flights were also used by a group from the High Altitude Observatory to overcome the brightness of the atmosphere in order to photograph the solar corona in white light. Their first attempt (Coronascope I) in 1960 was not very successful, but in their second try in 1966 with improved equipment (Coronascope II) their balloon reached an altitude of 99,000 ft (~ 30 km). From this height, they were able to obtain with their coronograph beautiful pictures of coronal streamers extending to several solar radii from the disc of the sun. Comparable pictures from the ground can be obtained only during the rare and fleeting moments of a total eclipse.

The infrared region extends from about 10,000 Å = 1 μ to approximately 1000 μ = 1 mm. Between the different absorption bands of H_2O and CO_2 there are several reasonably transparent windows in the 1 to 25 μ range through which some ground observations can be made. Infrared observations are now carried out also from special jet planes which can fly above most of the water vapor and carbon dioxide content of the earth's atmosphere, by reaching altitudes between 12 and 15 km. These recent jet observations have expanded tremendously the interest in infrared astronomy and they have been used, among other things, for the study of the infrared spectrum of the sun (Bijl, Kuiper, and Cruikshank, 1969). When normal telescopes are used for solar observations in the infrared, they are usually

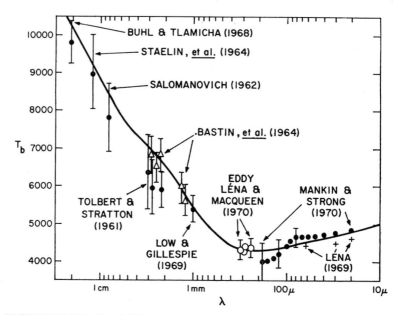

FIGURE 7.3-IV The brightness temperature of the sun vs. wavelength. The minimum temperature occurs in the lower chromosphere and is reached in the far infrared region (Linsky and Avrett, 1970)

shielded with a black polyethylene filter to prevent the visible and near infrared radiation from reaching the primary mirror of the telescope.

Infrared observations of the sun can provide very useful information about the minimum temperature region between the photosphere and the chromosphere. As seen from Figure 7.3-IV, the predicted minimum should occur in the spectral range near $100 \mu = 0.1$ mm where ground observations are impossible. The recent development of the Josephson detectors (junctions of two superconductors connected by a weak link) holds now a great promise for future submillimeter observations in space. These detectors are far superior to anything that was previously available in the submillimeter and millimeter range and have already been tried with great success in solar and planetary observations from the ground in the millimeter range (Ulrich 1969).

7.4. Solar Space Radioastronomy

Another spectral domain where space observations of the sun can be of great value is the long wavelength region of the radio spectrum. Ground measurements at these wavelengths are blocked by the terrestrial ionosphere which, with a daytime critical frequency between 5 and 10 MHz, sets the frequency limit for ground-based solar radio astronomy. Space vehicles, on the other hand, equipped with antennas and radio receivers and capable of flying above the maximum of the ionosphere ($h_{max} \simeq 300$ km), can extend our radio astronomical observations into the 10 to 0.1 MHz region of the spectrum. Space observations in this frequency range can provide extremely valuable information about the outer corona (Papagiannis, 1971).

All spaceborne radio antennas are practically omnidirectional because, like the small radio antennas of our cars, they are usually much shorter than the wavelength of the radio waves they receive, and thus they cannot focus on any particular region of the sky. In the hectometer range ($100 < \lambda < 1000$ m, or, $3 > f > 0.3$ MHz) the solar corona, with a temperature of the order of 10^{6}°K is less hot than the galactic background, which at frequencies around 1 MHz radiates like a black body at a temperature of about 10^{7}°K. The result is that with the low directivity antennas presently available for space radio astronomy (their beam width is considerably larger than the angular size of the solar corona), it is impossible to observe the quiet sun in the hectometer range. This is not the case, however, with the active sun because our space borne receivers can detect a good number of solar radio bursts which are intense enough to stand out above the galactic background.

Of special interest among the solar bursts that extend into the long wavelength region are the type II slow-drift radio bursts, which are caused by the advancing shock front from a flare event, and the type III fast-drift radio bursts, which are caused by a burst of nearly relativistic particles from the flare region streaming outwards through the solar corona. The instantaneous narrowband frequency of both type II and type III bursts corresponds to the local plasma frequency f_N (4.2-23) of the solar corona at the location of the disturbance. This is a generally accepted theory though other alternative theories have also been discussed in the literature (see Papagiannis, 1971).

As the disturbance which is exciting the radio burst moves outwards through the solar corona, the plasma frequency, which as seen from (2.3-12) is proportional to the square root of the local electron density N,

$$f_N = AN^{1/2} \tag{7.4-1}$$

decreases and the peak frequency of the burst drifts to lower frequencies. Let the electron density profile of the solar corona be given by an expression similar to the Baumbach-Allen formula (4.3-1),

$$N(r) = N_0 \left(\frac{r}{R_0} \right)^{-\mu} \tag{7.4-2}$$

and let V be the radial velocity of the advancing disturbance.

Let us also consider the general case when the velocity vector V makes an angle θ with the vector pointing toward the earth. The fequencies produced at the points O and A of Figure 7.4-I will arrive at the earth with a time delay dt. This delay is due to two reasons. One is that the radio noise produced at the point O must travel the extra distance $OB = OA \cos \theta$. The other is that the noise from the point A was produced later because the disturbance which excites the radio bursts had to first travel the distance OA with a velocity V. Hence, if we call $OA = dr$ we have,

$$dt = \frac{dr}{V} - \frac{dr \cos \theta}{c} = (1 - \beta \cos \theta) \frac{dr}{V} \tag{7.4-3}$$

where $\beta = V/c$. The second term of (7.4-3) can be neglected for the type II bursts where $\beta \ll 1$, but for the type III bursts, β is of the order of 0.2–0.6 and the second term of (7.4-3) must be retained. The angle θ can usually be determined from the heliocentric coordinates of the flare region which is associated with the radio burst.

The frequency drift df/dt in the vicinity of a certain frequency f, as we measure it from the earth, can be computed in terms of the above-mentioned parameters. Since $f = f_N$, we can use (7.4-1) and (7.4-2) to express f in terms of r. We can also express dt in terms of dr using (7.4-3), and thus df/dt represents a simple differentiation of an r-dependent function with respect to r,

$$\frac{df}{dt}\bigg|_s = \frac{V}{(1 - \beta \cos \theta)} \frac{df_N}{dr} = \frac{V}{(1 - \beta \cos \theta)} \frac{d(AN^{1/2})}{dr}$$

$$= \frac{VAN^{-1/2}}{2(1 - \beta \cos \theta)} \frac{dN}{dr} = \frac{VAN^{-1/2}}{2(1 - \beta \cos \theta)} \left(-\mu \frac{N}{r}\right)$$

$$= \frac{-\mu V}{2(1 - \beta \cos \theta)} \frac{AN^{1/2}}{r} = -\frac{\mu V f_N}{2(1 - \beta \cos \theta) r}$$

$$= -\left[\frac{\mu V}{2(1 - \beta \cos \theta) R_0 (A^2 N_0)^{1/\mu}}\right] f^{(2+\mu)/\mu} \qquad (7.4\text{-}4)$$

TOWARD THE EARTH

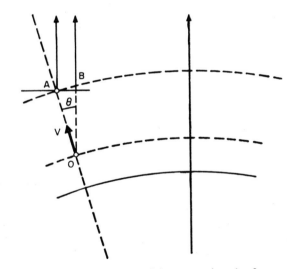

FIGURE 7.4-I   The geometry employed in computing the frequency drifts of the type-II and type-III solar radio bursts

Unfortunately (7.4-4) contains two unknowns for which we still have only approximate values. These unknowns are the velocity V of the exciter of the bursts and the electron density profile of the corona. In the case of the type III bursts, which are the ones that space radio receivers have been observing so far, the electron density is probably that of a coronal streamer which seems to be the pathway through which the exciter of the type III bursts propagates. The electron density of the streamers is believed to be

14*

about 10 times higher than the electron density of the surrounding corona, which is not known very well either at great distances from the sun. Thus (7.4-4) can yield the electron density profile of the streamer only if we assume a velocity for the exciter, and conversely the velocity of the exciter for an assumed profile of the streamer. Figure 7.4-II shows the average drift rate deduced by Hartz (1969) from a large number of type III bursts which he observed with the Alouette II satellite using sweep frequency radio receivers. More recent data from OGO-3 and OGO-5 indicate (Papagiannis, 1971) that this straight line continues at least down to 100 kHz, with df/dt remaining proportional to f^n, where $n \simeq 1.85$, which from (7.4-4) yields $\mu \simeq 2.35$ for (7.4-2).

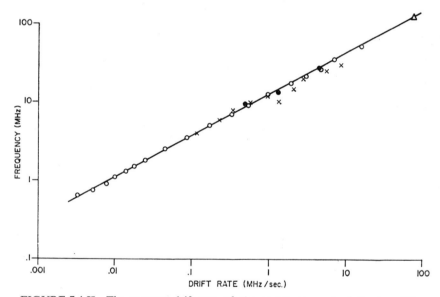

FIGURE 7.4-II   The average drift-rate of the commencement of the type-III solar radio bursts vs. frequency, deduced by Hartz (1969). This straight line extends to frequencies around 100 kHz and possibly even lower

The important contribution of space radioastronomical observations is that they allow us to follow the frequency drifts of the type III bursts into the hectometer and the kilometer range, i.e., up to a distance of 50–200 solar radii from the sun. In contrast, ground-based observations are limited by the terrestrial ionosphere to frequencies higher than about 10 MHz, i.e., to a radial distance in the corona of at most a few solar radii. Thus solar space radio astronomy offers a unique tool for the study of the outer corona. Unfortunately, due to the low resolution of our space-borne antennas we cannot follow the outward motion of the exciter and thus we cannot determine its speed in an independent way. People, however, have already considered the possibility of using two satellites as an interferrometer to solve this problem. This is the aim of the French project *Stereo*.

Fainberg and Stone (1970) have used a very smart statistical approach to determine the average velocity V of the exciters of the type III bursts. With the sensitive equipment and large antennas of the first Radio Astronomy Explorer (RAE-1) (see Section 8.6), they found that an active region of the sun emits continuously a large number (many thousands during one solar rotation) of type III bursts. This they call a *type III storm*. Following the rotation of such an active region across the solar disc, they were able to plot the average drift time between two neighboring frequencies vs. solar longitude. As seen from Figure 7.4-I, *OA* remains the same but *OB* changes as the angle θ changes and therefore df/dt changes with θ, as we can also see from (7.4-4). The relation between df/dt and θ provides an additional expression which, in conjunction with (7.4-4) allows us to determine V.

A typical example of their statistical analysis is shown in Figure 7.4-III. A least squares fit to their data has yielded an average velocity for the exciter

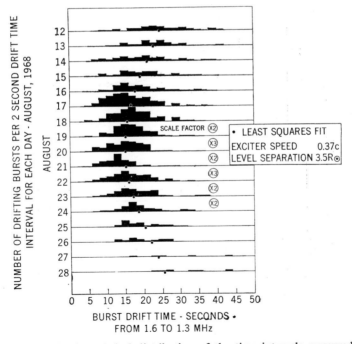

FIGURE 7.4-III A statistical distribution of the time intervals measured for the drifting of type-III radio bursts from 1.6 to 1.3 MHz. The minimum time occurs when the active region crosses the central meridian of the sun. A least squares fit to the data yields $V = 0.37$ c (Fainberg and Stone, 1970)

of $V = 0.35$–0.40 c. The spread in the drift times seen in Figure 7.4-III undoubtably represents some distribution in the drift velocities of the exciter. It is believed, however, that it reflects more the inhomogeneities of the electron density in the different filaments of the coronal streamer through which the individual exciters happen to pass. A very interesting byproduct of Fainberg and Stone's analysis was that the minimum of the drift-time plots, like the one in Figure 7.4-III, occurs a day or two later for lower frequency intervals. The explanation seems to be that the exciters follow curved paths, most probably as a result of the Archimedian spiral structure of the interplanetary magnetic field (see Section 5.5), and therefore points at larger radial distances (lower plasma frequencies) cross the central meridian ($\theta = 0$) with a certain delay.

Figure 7.4-IV shows the drift, rise and decay times for a typical type III burst as it was recorded with the discrete frequency and the sweep frequency radiometers of the RAE-1 satellite. The decay times of the type III bursts

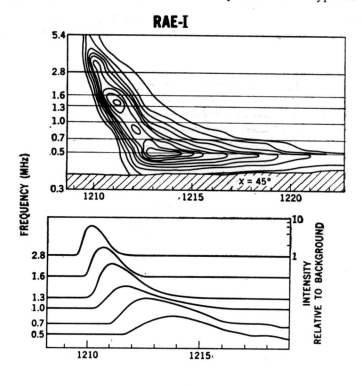

FIGURE 7.4-IV   A typical type-III solar radio burst recorded by both the sweep frequency (upper diagram) and the discrete frequency (lower diagram) receivers of the RAE-I satellite (Fainberg and Stone, 1970)

at different frequencies can be used in principle to determine the temperature in the region of the solar corona where the burst at this frequency was produced. In this way one can obtain not only an electron density profile but also a temperature profile of the outer corona.

Assuming that the decay of the disturbance which produces the type III bursts is achieved through collisional damping, we can relate the time constant τ_0 (the time interval in which the intensity of the radio burst decreases by a factor of $1/e$) to the local collision frequency of the corona. Since the solar corona is a fully ionized plasma where $\nu = \nu_i$, from equations (2.4-3) and (2.4-5) we obtain,

$$\tau_0 = \frac{1}{\nu_i} \simeq \frac{1}{50NT^{-3/2}} \qquad (7.4-5)$$

where T is the local temperature of the corona in °K and N the local electron density in el/cm³. Since N is related to f_N (4.2-3) and f_N is equal to the frequency f of the burst, we can express the temperature T of (7.4-5) in terms of τ_0 and f. The result is,

$$T \simeq 7 \times 10^{-5}\tau_0^{2/3}f^{4/3} \qquad (7.4-6)$$

where τ_0 is in seconds, f in Hz, and T in °K.

The application of (7.4-6) to actual data has produced values of the coronal temperature which, near the sun, are in good agreement with generally accepted values. At longer distances, however, they tend to fall much faster with distance than expected. It has been suggested that the coronal streamers are denser but cooler than the ambient corona so as to balance the pressure inside and outside the streamers. This could account for some, but not for all of this discrepancy. Another possibility is that at large distances from the sun, where the collision frequency becomes very low, collisional damping is replaced by hydromagnetic damping. This is a very interesting subject which has not yet been carefully investigated.

Type III solar radio bursts in the hectometer range have been observed and analyzed by several groups around the world. No other types of solar radio bursts have been detected as yet in the hectometer range with any high degree of certainty. The intensity of the type III bursts seems to increase with decreasing frequency. Type III bursts have a wide range of intensities. Some stand above the integrated galactic background by as much as 60 db, while others are barely detectable above the background noise level. It seems also that the number of type III bursts increases tremendously at the lower intensities.

It appears also that the intensity of a type III burst might have ups and downs with frequency. Most of the type III bursts recorded in the hectometer range from space vehicles are actually a continuation of the type III bursts observed in the meter range from the ground. This, however, is not always the case and there have been several instances when type III bursts were observed in the one frequency range and not in the other. This is probably the result of electron density inhomogeneities along the coronal streamers which affect the intensity and the propagation characteristics of the type III solar radio bursts in the different frequencies of the radio spectrum.

7.5. Planetary Space Astronomy

Planetary space astronomy is somewhere in the middle between ground-based planetary astronomy and the exploration of the planets with space probes (see Appendix II). As in all such cases it combines many of the advantages, and of course some of the disadvantages of the other two. Planetary space astronomy is more expensive than ground astronomy but far less costly than the direct exploration of the planets. Orbiting observatories around the different planets can do in principle all that planetary space astronomy expects to do. These missions, however, are much more difficult and expensive, the telemetry of the data from such long distances becomes a serious problem, and it does not seem very likely that we will have an orbiting satellite around Jupiter, for example, before the 80's or the 90's. On the other hand, with the opportunity to conduct observations over the entire spectrum, planetary space astronomy has a tremendous advantage over ground-based astronomy.

Planetary observations from space-borne observatories around the earth also offer the opportunity for long-term projects, which require a much longer observing period than we can get with single fly-by missions to the planets. Such continuous observations over a wide spectral range will allow us to study the long-period (seasonal, etc.) variations in the atmosphere, the albedo, and the ground temperature of the planets and conceivably might reward our patience with an occasional unexpected event such as a volcanic eruption. Furthermore the ultraviolet and infrared observations of planetary space astronomy will be able to provide the correct specifications for the ultraviolet and the infrared spectrometers to be carried by the preciously few space probes we will be able to send to the planets in the coming years.

The region of interest of planetary space astronomy extends from the far ultraviolet to the far infrared, but at least one of the planets, namely Jupiter, is also an object of great interest in the long wavelength region of the radio spectrum.

Ultraviolet observations of the planets can provide very useful information on the constitution of their atmospheres. The change of the planet's albedo with wavelength allows us to evaluate the relative importance of the atmosphere, the clouds and even the surface of the planet in absorbing and reflecting the sunlight. From such measurements we can draw conclusions about their consistencies, densities, etc. Furthermore, by searching for broad absorption bands we might be able to identify some major atmospheric constituents such as oxygen, or some strongly absorbing molecules, such as ozone. We have already mentioned that the terrestrial atmosphere becomes opaque to the ultraviolet radiation at 3,000 Å because of the ozone absorption.

From the emission spectrum of a planet in the ultraviolet domain we might be able to detect a large number of major and minor atmospheric constituents in their molecular, atomic, or ionized forms. This includes the emission lines of atomic oxygen at 1,304 and 1,356 Å, the lines of atomic nitrogen at 1,200, 1,493, and 1,744 Å, the emission bands of neutral and ionized molecular nitrogen, the emission bands of nitric oxide, etc. All these lines and bands are excited in the upper atmosphere of the planets by resonance and fluorescence scattering of the solar ultraviolet radiation and to some extent by photoelectron impact excitation.

Most of these weak emission features occur in the 1,300 to 1,800 Å range and their detection is considered possible only because, in this spectral range, the planets are practically black due to the strong absorption by many of their atmospheric constituents. Because the intensities of these emission features are very low, long integration times (several hours) and rather large space telescopes ($>20''$) are required. As a result such observations cannot be carried out from rocket flights. There are, however, many other observations that can be achieved with rockets, and, as a matter of fact, until the first planetary observations with the *Orbiting Astronomical Observatory* (OAO) in 1969, all our previous ultraviolet data about the planets had been obtained with rocket flights of short duration.

One of the most interesting features of the ultraviolet spectrum is the Lyman-alpha line of atomic hydrogen (1,216 Å). By studying the emission of this line we can compute the exospheric temperature which is of major importance in reconstructing the past history of the atmosphere of a planet. Also of great significance is to determine the relative abundance of hydrogen and helium in the atmospheres of the Jovian planets. Finally by studying with high angular resolution the different spectral features of planetary atmospheres across the limb of the planets we can obtain models of their atmospheres. All these data about planetary atmospheres will undoubtably include some seasonal and solar cycle variations, which planetary space astronomy can study very effectively.

In the visible region, telescopes with apertures in the 20- to 60-inch range will provide angular resolutions of the order of 0.1 arc second. Since the angular size of Venus, Mars, Jupiter, and Saturn are all of the order of 20 arc seconds, a resolution of 0.1 arc second will allow us to study in considerable detail the different features of these planets. It might also be possible to detect the existence of craters on the larger moons of Jupiter. Furthermore, long-term observations of cloud formations and their motions over the disc of the planet can become the first steps of *planetary meteorology*. It is interesting to note that one of the objectives in a repeat flight of Stratoscope II in 1970 was to obtain with its 36-inch balloon borne telescope photographs of the planet Uranus to test the theory that Uranus, unlike Jupiter and Saturn, has no clouds.

Infrared observations of the planets are also of great importance because they provide data on the abundances of molecules such as H_2O, CH_4, NH_3, CO_2, etc. in the atmospheres of the planets. They also allow us to study the temperature profile of the atmosphere by measuring the brightness temperature of a planet at different wavelengths in the far infrared and the microwave region. The recent development of the Josephson detector (Ulrich, 1970) opens new possibilities in this area. Finally, with the use of larger space telescopes we can also study the latitude variations in the temperature of a planet.

Observations in the infrared region, where the thermal emission of the planets occurs, coupled with the accurate measurements of albedos over the entire range of the solar spectrum, will allow us to determine the heat budget of a planet. Such studies will make it possible to answer the very fundamental question of whether the major planets still possess an internal heat source. Preliminary results seem to suggest that this is true in the case of Jupiter. Since the temperature and possibly even the albedo of a planet undergoes seasonal and solar cycle variations, one needs the longterm observing capability of planetary space astronomy to answer this question accurately.

Jupiter is also a very interesting planet for space radio astronomy. In 1954, Burke and Franklin observed the emission of extremely intense decametric radio bursts from Jupiter. The spectrum of these bursts peaks around 20 MHz but falls very steeply at the higher frequencies, so that the Jovian bursts can hardly be detected above 40 MHz. The spectral character of the bursts at lower frequencies is still unknown because the terrestrial ionosphere cuts off ground observations at these frequencies. For this reason, space radio observations of the dynamic spectra of the Jovian bursts are of great importance in our effort to understand the complex magnetohydrodynamic interactions that take place in the ionosphere and magnetosphere of Jupiter.

This problem has become even more interesting since it was discovered that Io, one of Jupiter's moons, exercises a strong control on the decametric emission of Jupiter (Warwick, 1968). Careful observations have shown that the frequency of occurrence and the intensity of Jupiter's radio bursts increase unmistakeably when Io is at a certain configuration relative to the earth–Jupiter line. It seems that a hydrodynamic wave, which is excited by the passage of Io, focusses under certain conditions into a coherent burst in a certain preferential direction. Of interest will also be the search for any effects in the vicinity of 100–200 kHz, where the gyrofrequency at the orbiting distance (~ 6 Jovian radii) of Io is believed to occur.

In this section we should also include the observation of comets from space observatories. Ultraviolet observations are again of prime importance because they can reveal the chemical composition of the comets. Observations of the Tago-Sato-Kosaka and of the Bennet comets in 1969 with the ultraviolet detectors of the second Orbiting Astronomical Observatory (OAO-2), have revealed that the comets are surrounded by a large hydrogen cloud. These observations have also shown that the hydroxyl radical (OH) is several hundred times more abundant than cyanogen (CN). These results suggest that one of the main constituents of comets is water in the form of ice which, as the comet approaches the sun, melts, evaporates, and is partially dissociated by the ultraviolat radiation of the sun.

7.6 Bibliography

A. Books for Further Study

1. *American Space Exploration, the First Decade*, W. R. Shelton, Little, Brown and Co., Boston, Mass., 1967.
2. *History of Rocketry and Space Travel*, W. von Braun and F. I. Ordway III, (Revised edition) Thomas Y. Crowell Co., New York, N. Y., 1969.
3. *Telescopes in Space*, Z. Kopal, Faber and Faber, London, England, 1969.
4. *Les Observatoires Spatiaux*, J.-C. Pecker, Presses Universitaires de France, Paris, France, 1969.
5. *Electromagnetic Radiation in Space*, ed. by J. G. Emming, Springer-Verlag, New York, N. Y., 1967.
6. *Solar Physics*, ed. by J. N. Xanthakis, John Wiley and Sons, New York, N. Y., 1967.
7. *Infrared Astronomy*, ed. by P. J. Brancazio and A. G. W. Cameron, Gordon and Breach, New York, N. Y., 1968.
8. *Significant Achievements in Space Astronomy* 1958–1964, NASA SP-91, U. S. Gov. Print. Office, Washington, D. C., 1966.
9. *Astronomy in Space*, H. Newell, H. Smith, N. Roman, and G. Mueller, NASA SP-127, U. S. Gov. Print. Office, Washington, D. C. 1967.
10. *Manned Laboratories in Space*, ed. by S. F. Singer, Springer-Verlag, New York, N. Y., 1969.

11. *Planetary Exploration* 1968–1975, Space Science Board, NAS/NRC, Washington, D. C., 1968.
12. *A Long-Range Program in Space Astronomy*, ed. by R. O. Doyle, NASA SP-213, U. S. Gov. Print. Office, Washington, D. C., 1969.
13. *TRW Space Log*, A quarterly publication of the TRW Systems Group, Redondo Beach, Calif., Vol. 9, No. 4, 1970.
14. *High Energy Astrophysics*, T. C. Weeks, Chapman and Hall, London, England, 1969.

B. Articles in Scientific Journals

Burke, B. F. and K. L. Franklin, Observations of a variable radio source associated with the planet Jupiter, *J. Geophys. Res.*, **60**, 213, 1955.
Dolan, J. F. and G. G. Fazio, The gamma-ray spectrum of the sun, *Rev. of Geophys.*, **3**, 319, 1965.
Fainberg, F. and R. G. Stone, Type III solar radio burst storms observed at low frequencies, *Solar Physics*, **15**, 433, 1970.
Goldberg, L., Ultraviolet Astronomy, *Sci. Amer.*, **220**, No. 6, 92, 1969.
Hartz, T. R., Type III solar radio noise bursts at hectometer wavelengths. *Plan. Space Sci.*, **17**, 267, 1969.
Linsky, J. L. and E. H. Avrett, The solar H and K lines, *Publ. Astron. Soc. Pac.*, **82**, 169, 1970.
Neupert, W. M., X-rays from the sun, *Ann. Rev. Astron. Astrophys.*, **7**, 121, 1969.
Papagiannis, M. D., Space Astronomy, *Science*, **166**, 775, 1969.
Papagiannis, M. D., Studies of the outer corona through space radio astronomy, *Physics of the Solar Corona*, ed. by C. Macris, D. Reidel Publ. Co., Dordrecht-Holland, 1971.
Spitzer, L., Jr., Astronomical research with large space telescopes, *Science*, **161**, 225, 1968.
Strauss, F. M. and M. D. Papagiannis, A model for the source of solar flare x-rays, *Astrophys. J.*, **164**, 369, 1971.
Ulrich, B. T., Millimeter-wavelength observations of solar activity, the moon, and Venus with a Josephson junction detector, *Bulletin AAS*, **2**, 222, 1970.
Underwood, T. G., Solar x-rays, *Science*, **159**, 383, 1968.
Vaiana, *et al.*, X-ray structures of the sun during the important 1 N flare of 8 June 1968, *Science*, **161**, 564, 1968.
Warwick, J. W., Radiophysics of Jupiter, *Space Sci. Rev.*, **6**, 841, 1967.

CHAPTER 8

GALACTIC SPACE ASTRONOMY

8.1 Introduction

In the introduction of Chapter 7 we reviewed the spectral domains, the basic problems, and the general objectives of space astronomy, i.e., of astronomical observations conducted above the obstructing regions of the earth's atmosphere. We also discussed its application to the study of the sun and the planets, which had been the subjects of some of the previous chapters of the book. In this final chapter we will discuss the prospects and the accomplishments of galactic space astronomy. This is a topic much closer to astronomy, with which some of our readers might be less familiar. For this reason we will preface this chapter with a brief, but hopefully comprehensive, introduction to the basic concepts of galactic astronomy.

The effective temperatures of the stars range from lower than 3000°K to higher than 30,000°K. Astronomers have classified the stars in different spectral types, i.e., essentially in different temperature groups because the spectrum changes as the temperature changes. A special letter has been assigned to each spectral class and the seven basic classes O–B–A–F–G–K–M form a temperature as well as color sequence from the very hot, blue, early type O and B stars, to the relatively cold, red, late type K and M stars. Each spectral class is further divided into ten numbers, 0 to 9, as we progress from hotter to colder.

Astronomers scale the *brightness b* of the different stars by assigning an *apparent magnitude m* to each star. Stellar magnitudes are proportional to the logarithms of stellar brightnesses. The basic reason for this relation is our physiological reaction to optical stimuli, which is proportional to the logarithm of their intensity (Pogson's rule). By assigning zero magnitude to a star of certain brightness, we have a magnitude scale given by the expression,

$$\frac{2}{5}(m_1 - m_2) = \log_{10}\left(\frac{b_2}{b_1}\right) \tag{8.1-1}$$

From (8.1-1) we see that higher magnitudes correspond to fainter stars, so that if, e.g., $m_1 = 15$ and $m_2 = 10$, we have that $b_2 = 100b_1$. The magnitude of Sirius, which is the brightest star in the sky is $m = -1.5$, while the weakest light sources that can be seen with the 200-inchMt.Palomar telescope have magnitudes between 23 and 24.

The apparent magnitude m of a star depends on the intrinsic luminosity of the star, its distance (the brightness of a light source decreases as the square of the distance), and the losses the light of a star suffers on its way to us due to absorption and scattering by the tiny dust particles of the interstellar space. This *interstellar extinction* is stronger for the blue light than for the red light, because red light with its longer wavelength bypasses the tiny dust grains easier. Thus, interstellar absorption produces *interstellar reddening*, i.e., it makes a star appear redder than it ought to be according to its temperature, which we can determine from spectroscopic observations. A similar effect happens when the color of our sun changes to red at sunrise and sunset. As the sun approaches the horizon, the solar rays travel a longer path through the earth's atmosphere losing as a result more of the blue component of the solar light. The relative absorption at the different wavelengths allows us in principle to study the density and the physical characteristics (size, composition, etc.) of the interstellar dust.

When the apparent magnitude of a star is corrected for absorption and distance effects, we obtain the *absolute magnitude M* of the star. This is the magnitude the star would have without any absorption at a distance of 10 parsecs. The *parsec* is a unit of length (1 pc = 3.1×10^{18} cm) commonly used in astronomy. So also is the *light year* (1 l.y. = 9.5×10^{17} cm) which is the distance light travels in one year. The relation of these two units is: 1 pc = 3.26 l.y. Larger distances are measured in kiloparsecs (kpc) and megaparsecs (Mpc) or in thousands, millions, and billions of light years.

By plotting the absolute magnitude against the spectral class of a large number of stars, we obtain a diagram (Figure 8.1-I) where most of the stars fall along a diagonal line. This is called the *Hertzsprung-Russell diagram* and the diagonal line is called the *main sequence*. In a sense it is similar to plotting number of people vs. height. As one can guess, such a statistical study will show that there is a much larger number of people in the 5–6 ft. bracket than in the 2–3 ft. bracket, simply because people spend most of their years at the adult height while passing quickly through the shorter heights of childhood. In a similar manner, the fact that most stars fall on the main sequence means that this is where they spend most of their lives.

After a star is formed from a dense cloud of gas and dust, it takes itsplace on the main sequence, its exact position (how hot, or how bright) depending essentially on the mass of the star. Stars remain there for most of their active lives, glowing away steadily the energy which is generated in their cores by the "burning" of hydrogen into helium. This burning proceeds much faster in the most massive stars which, as a result, are hotter and stay for a much shorter period in the main sequence. When all the hydrogen in the core of the star has been converted into helium, the star departs from the main sequence.

Hydrogen burning occurs now only in a relatively thin shell around the helium core which begins to contract. Approximately half of the gravitational energy released by this contraction is kept in the core producing a large increase in its temperature. The other half is released to the hydrogen burning shell which starts producing more energy, and to the outer layers of the star which are forced to expand to gigantic proportions. This expansion lowers the temperature of the outermost layers and as a result changes the color of the star to red.

These *red giants*, or *supergiants*, with diameters ten, one hundred, or even one thousand times their initial diameters, occupy the upper right corner of the H–R diagram. When the contracting core reaches a temperature of about 10^8°K, helium starts burning rapidly to carbon ($3\,\mathrm{He} \rightarrow \mathrm{C}$), and through further nuclear reactions to other elements of intermediate atomic number, such as oxygen, neon, magnesium, etc. This new source of energy produces drastic changes in the star which shrinks, becomes hotter, and moves along a nearly horizontal trach toward the upper central and upper left regions of the H–R diagram.

The evolution of the different stars beyond the red giant stage depends greatly on their initial mass and it is not yet very well understood. In their

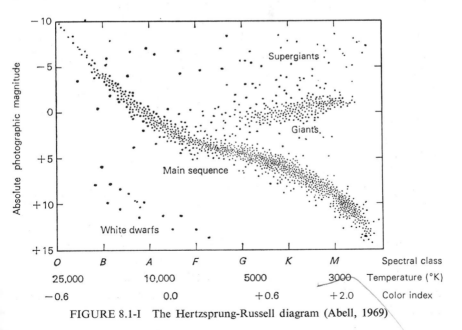

FIGURE 8.1-I   The Hertzsprung-Russell diagram (Abell, 1969)

post-giant evolution, stars usually go through unstable conditions, some-times becoming variable stars (Cepheids, RR-Lyrae, etc.) with periods ranging from a few hours to several months. It is also quite certain that many stars go through stages in which they eject a substantial part of their matter into the interstellar space (shell stars, Wolf-Rayet stars, planetary nebulae, novae and supernovae).

The final stage in the life of most stars seems to be the *dwarf* stage. After the stars have used up as much as possible of their nuclear fuel in a sequence of burning processes (hydrogen, helium, carbon, etc.) they finally contract to a density of 10^3–10^8 gr/cm^3, depending on the degree of *degeneracy* of their matter. *White dwarfs* are dwarf stars that are still hot from the gravitational energy released during this final contraction. Theoretical studies have shown that white dwarfs cannot have a mass larger than about 1.2 times the mass of the sun. It seems therefore quite certain, even from a theoretical point of view, that many stars will have to eject a substantial part of their mass if they are to reach the white dwarf stage.

In the case of a *supernova*, the ejection of stellar matter takes the form of a catastrophic explosion. It is estimated that supernovae occur only once every hundred years or so in each galaxy, but they are so bright that many of them have been detected even in distant galaxies. The total energy released in a supernova is of the order of 10^{49} ergs, while a substantial fraction of the total mass of the star is ejected into the interstellar space at speeds of the order of 10^9 cm/sec. The light curve of a supernova usually shows an exponential decay with a half life of approximately 55 days.

Our understanding of what causes a supernova is still not entirely clear. It seems that at the very late stages of nuclear burning (formation of iron at temperatures around 10^9°K), the cores of certain stars suffer a gravitational collapse, collapsing probably into a *neutron star* with a density of 10^{14}–10^{16} gr/cm^3. The matter of neutron stars consists primarily of neutrons which were produced by the fusion of protons and electrons under the tremendous pressure which is generated during the gravitational collapse. A powerful shock wave, which is produced from the collapse of the core, bounces outwards and expels the outer layers of the star into the surrounding space. The ejected matter forms a rapidly expanding nebula around the core star.

A *nova* is a stellar explosion, similar in appearance to a supernova, but of a much smaller magnitude and probably of a very different cause. Since many and perhaps all novae are members of close binaries, it is possible that this close association of two stars triggers the eruption of a nova at some stage of their evolution. There are several stars which are known to be recurring novae.

In the year 1054 A.D. Chinese astronomers recorded a spectacular super-nova which produced a very bright star that was visible even during the day. This has now been identified with the *Crab-Nebula*, a luminous cloud in the constellation of Taurus, which is still expanding with a velocity of approximately 1,300 km/sec. The Crab Nebula is at a distance of only about 4,000 light years from us and has become a real treasure-house for Astronomy. Thus in 1949 it was the first radio source (*Taurus A*) to be identified with a visible object beyond our solar system, and in 1964 it became the first x-ray source (*Tau XR-1*) to be identified with a visible object.

The detection of the first *pulsar* by Hewish *et al.* (1967) was a great new discovery for Astronomy. Pulsars emit radio pulses, typically at the rate of one per second. The Crab Nebula again played a very prominent role because the star at the center of the nebula, which astronomers had suspected for a long time as the supernova remnant, turned out to be a pulsar. Among the more than 50 pulsars known, the Crab Nebula (NP 0532) has the fastest pulsing rate (about 30 times per second) and it is the only one which simultaneously with the radio pulses emits also optical and x-ray pulses.

It is now generally accepted that pulsars are rapidly rotating neutron stars, with a mass similar to that of the sun, a diameter of about 10 miles, and very probably an enormous magnetic field. When it was discovered that the pulsing rate of pulsars is actually slowing down, the Crab Nebula pulsar was again on the top of the list. This discovery, as we will see in Section 8.3, has provided not only very important information about the nature of pulsars and the whole emission process of the Crab Nebula, but also a possible clue for the production of cosmic rays.

Stars are usually grouped together in billions forming a galaxy. Our own galaxy, the *Milky Way*, contains approximately 10^{11} stars. Most of these stars are found near a plane, which is called the *galactic plane*, and form a disc which has a diameter of nearly 100,000 light years and an average thickness of only about 2,000 light years (Figure 8.1-II). The galactic disc forms a bulge near the center of the disc which contains the *nucleus* of the galaxy. In the nucleus, stars and interstellar matter are packed at much higher densities than in the rest of the galactic disc. Our sun is located in the plane of the galaxy at a distance of approximately 30,000 l.y. from the galactic center, which is in the direction of the constellation of Sagit-tarius. The nucleus of our galaxy is a very intriguing region and it is a strong emitter of radio and infrared radiation.

The stars and the interstellar matter of our galaxy are concentrated in several spiral arms which start near the nucleus of the galaxy and wind outwards toward the rim of the galactic disc (Figure 8.1-II). The entire disc of the galaxy rotates about the galactic center, but not exactly as a solid body. At the distance of our sun it is estimated that a complete revolution takes approximately 250 million years. Most of our knowledge about the structure of our galaxy comes from radioastronomical observations. The thinly distributed interstellar dust and the occasional denser dust clouds that are found in the interstellar space limit our visual observations in the galactic plane to a distance of only about 1 kpc.

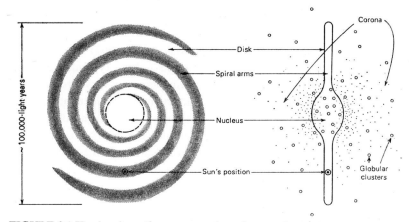

FIGURE 8.1-II   A schematic representation of our galaxy with a top view at the left and an edge-on view at the right (Abell, 1969)

The interstellar space also contains a significant amount of gas, mostly neutral hydrogen, which is detected by its characteristic emission (or absorption) at the radio wavelength of 21 cm. The neutral hydrogen detected from its 21-cm line is actually atomic hydrogen. We know very little about molecular hydrogen in the interstellar space, and our only direct information has been a recent detection of molecular hydrogen lines in the ultraviolet from a space vehicle. In recent years we have also identified helium and we have been able, through radio spectroscopy, to detect several other more complex chemical compounds such as the hydroxyl radical (OH) at 18 cm, formaldehyde (H_2CO) at 6.21 cm, ammonia (NH_3) at 1.25 cm, hydrogen cyanide (HCN) at 2.8 mm, etc. Extra-long base interferometric observations (e.g., one radiotelescope in Europe and one in America synchronized with atomic clocks to work together as an interferometer) have shown that certain regions where some of these compounds have been detected are exceedingly small (of the order of 10 Astronomical Units) and probably are stars in formation.

The interstellar space is permeated by a weak magnetic field ($\sim 10^{-6}$ Gauss) and also contains cosmic rays which spiral around the lines of force of this weak galactic field. A sparse "haze" of individual stars and a good number of *globular clusters* (tightly-bound groups of many thousands of stars) are found on both sides of the galactic plane (Figure 8.1-II) forming a spherical, or better ellipsoidal region which is called the *galactic halo* or the *galactic corona*. It is still a matter of debate whether the galactic halo also has a very weak magnetic field and some very tenuous interstellar matter. It is expected that space astronomy will be able ultimately to provide the answer to this question.

Our galaxy is a rather typical spiral galaxy of which there are many in the universe. Actually it is quite similar to the *Andromeda galaxy* (M31) which, at a distance of about 2.2 million l.y., is one of our nearest galaxies. There are also galaxies of no clear shape. These are called *irregular galaxies* with typical examples the *large* and the *small Magellanic Clouds* which are also members of our small (~ 20) *local group* of galaxies. A different type of galaxy is that which is simply a spherical or ellipsoidal system of stars with no indication of any spiral structure. These are called *elliptical galaxies*. Elliptical galaxies as a rule are extremely weak emitters of radio radiation, while spiral galaxies often emit a detectable flux in the radio domain of about 10^{38} ergs/sec. This radio emission is mostly synchrotron radiation produced by relativistic electrons, of primary or secondary cosmic ray origin, gyrating in the weak magnetic fields of these galaxies.

There are, however, also some galaxies which emit up to a million times more energy in the radio domain. These are called *radio galaxies. Cygnus A,* with a radio emission of about 10^{44} erg/sec is one of the strongest ones and has been identified with a pair of galaxies which are either colliding or being separated from a common mother galaxy. *Virgo A,* another strong extragalactic radio source, has been identified with the giant elliptical galaxy M87 which has a long jet streaming out of its galactic center. This jet, which is also a strong emitter of polarized synchrotron radiation, gives a clear indication that M87 has undergone a recent violent explosion. Some bright radio sources have been identified also with *Seyfert galaxies*, which are characterized by their extremely bright, star-like nuclei.

In most of the radio galaxies we have found that the radio emission comes from two regions that are usually symmetrically located on either side of the optical galaxy and often at a distance much larger than the diameter of the galaxy. This is a rather strong indication that the matter which is now emitting the radio radiation was ejected from the galaxy following a major galactic explosion. It is interesting to note that pictures of the irregular galaxy M82 indicate that such an explosion is taking place in this galaxy as we see it now. In Section 8.3 we will see that at least one of the strong radio galaxies (M87) is also an x-ray source, and in Section 8.4 we will

15*

see that many of these peculiar galaxies are also extremely strong emitters of infrared radiation.

In 1960 a new category of radio sources was discovered which were named *Quasi-Stellar Objects* (QSO), or for short *Quasars*, because their optical counterparts were small, star-like objects. In 1963 Maarten Schmidt realized that the mysterious spectral lines in one of the quasars (3C 273) were simply some of the Balmer lines of hydrogen and a line of ionized magnesium which had been Doppler-shifted from the ultraviolet region, where they normally are, into the visible region of the spectrum. This shifting of spectral lines toward longer wavelengths is called *red shift* and is usually the result of a high recession velocity between the object that emits the spectrum and the observer.

If V is the recession velocity and c the speed of light, the red shifted wavelength λ' is related to the original wavelength λ through the expression,

$$\frac{\Delta\lambda}{\lambda} = \frac{\lambda' - \lambda}{\lambda} = \left(\frac{1 + V/c}{1 - V/c}\right)^{1/2} - 1 \qquad (8.1\text{-}2)$$

In the non-relativistic case, where $V \ll c$, (8.1-2) reduces to the much simpler form,

$$\frac{\Delta\lambda}{\lambda} = \frac{V}{c} \qquad (8.1\text{-}3)$$

Previous observations had shown that all distant galaxies display a red shift and that their recession velocity V is proportional to their distance r, i.e.,

$$V = Hr \qquad (8.1\text{-}4)$$

where $H \simeq 80$ km/sec per megaparsec $\simeq 2.6 \times 10^{-18}$ sec^{-1}, is the famous *Hubble constant*.

This seems to be a general property of the universe (*expanding universe*) which optical observations have verified up to a distance of about 2 billion l.y. Quasars, however, display in general extremely large red shifts with a strong concentration around $\Delta\lambda/\lambda = 1.95$ (Burbidge and Birbidge, 1969). For $\Delta\lambda/\lambda = 2$ we find from (8.1-2) that $V = 0.8\,c$ which, extrapolating from smaller recession velocities, implies that some of these quasars are at a distance of 8–9 billion l.y. This makes them the farthest, and as a result the brightest known objects in the universe. It should be mentioned that some astronomers have tried to avoid this cosmological explanation by suggesting that the red shifts might be due to strong gravitational fields, or that the quasars have been ejected at these high velocities from some nearby exploding galaxies. These alternative theories, however, result in even more formidable problems and thus they have not gained much support.

Both the optical and the radio intensity of quasars fluctuate irregularly, changing by a factor of 2 or more in periods often shorter than one month. This means that a tremendous amount of energy is released in a volume less than one light-month in diameter. Quasars as a rule are very blue objects, a property which has allowed Sandage to discover many quasistellar objects with very large red shifts that are not radio sources. Frank Low, on the other hand, has found that several quasars emit extremely large amounts of radiation in the far infrared, and 3 C 273 has been found to be an x-ray source too. 3 C 273, by the way, also has a long jet streaming away from the nucleus of the source and is probably an indication, as in the case of M 87, of the explosive nature of these objects. It is possible that quasars, Seyfert galaxies, radio galaxies, and normal galaxies like ours represent some kind of a sequence in galactic evolution, but this is a subject which we are just now starting to uncover. It is hoped that space astronomy will provide many new clues in our effort to understand some of these puzzling problems of galactic evolution.

Galaxies tend to form groups or clusters which have from a few members, like our own local group, to a thousand members or more, like the *Virgo cluster* of galaxies. Some astronomers believe that our local group, as well as the Virgo cluster, belong to a larger system of galaxies which occupies a volume 100 to 150 million l.y. in diameter, and which they call the *supergalaxy*. It is suggested that there might be many such superclusters of galaxies in the universe, but beyond this order of grouping, the universe appears to be uniform in all directions, so that on a sufficiently large scale our universe can be considered as homogeneous.

As we have mentioned above, we have clear evidence from the red shifts of distant galactic clusters that our universe is expanding, and that the expansion is uniform in all directions. This does not necessarily imply that our galaxy is at the center of the universe. In a two-dimensional analogy we can think of dots on the surface of an expanding balloon where all dots run away from each other and no dot occupies a position which is more central than any other one. The proponents of the *big-bang theory*, currently the leading cosmological theory, believe that all the universe was set on its present expanding course by an explosion (big-bang) which took place 10–20 billion years ago. At that time, according to this theory, all the matter of the universe was packed in the form of photons and elementary particles in a superdense fireball which exploded sending matter running away in all directions.

There are several astronomical observations which concur on the above-mentioned age of the universe (ages of stellar clusters, the Hubble constant, radioactive dating, etc.), and recently the big-bang theory has received additional support with discovery of the *3°K background radiation* (Penzias and Wilson, 1965). This is a uniform radio emission which has the same intensity in all directions and its spectrum is that of a black body at a temperature of about 3°K. Both Gamow and Dicke had predicted this isotropic radio background claiming that this is the radiation from the primeval fireball after it has been weakened by the continuous expansion of the universe.

A very intriguing question is whether our universe will continue to expand forever. We could answer this question if we could determine the present average density of the universe. The critical density that will produce a *closed universe*, i.e., a universe in which the present expansion will ultimately stop and will be followed by a contraction, can be estimated, to a first approximation, by equating the kinetic energy and the gravitational potential of a body of mass m moving radially away with a velocity V from a large sphere of radius r and density ϱ. Thus we have,

$$\frac{1}{2}mV^2 = G\left(\frac{4}{3}\pi r^3 \varrho\right)\frac{m}{r}$$

(8.1-5)

which gives the relation,

$$V = \left(\frac{8}{3}\pi G \varrho\right)^{1/2} r$$

(8.1-6)

But (8.1-6) has exactly the same form as (8.1-4) and thus by equating $(\frac{8}{3}\pi G\varrho)^{1/2}$ to the known value of H we can compute the critical density, which is found to be of the order of $1-2 \times 10^{-29}$ gr/cm. If the average density of the universe is less than this critical limit, then we have an *open universe* which will continue to expand indefinitely.

Optical observations tend to favor the open model because the visible matter incorporated into the galaxies yields a much smaller average density. The existence in the galaxies of a large number of *black holes* (massive stars, denser than neutron stars, from which light cannot escape due to their very high gravitational field) which has been suggested by some theoreticians, is still an untested and rather doubtful hypothesis. As we will see in Section 8.3, however, recent x-ray observations might have finally detected the *missing matter* that would "close" the universe in the form of a rarefied hot plasma occupying the vast spaces between the galaxies.

The contraction stage in a closed model might ultimately lead to a new big-bang and thus possibly to an *oscillating universe* which would go through periodic phases of expansion and contraction. Some of the evolutionary cosmological theories believe that the universe is composed of equal amounts of matter and antimatter, which remain separated through a not very well

understood mechanism. In these theories the contracting stage in an oscillating universe will be reversed when the annihilation of matter and antimatter in a denser universe will reach a sufficiently high level.

Competing with all the evolutionary cosmologies is the *steady state*
theory, which says that the universe has always been and will always be
the same. This implies that there was no beginning of any kind and there
will be no end. This theory is forced to postulate a continuous creation
of matter so that the universe can maintain the same density in spite of
its expansion. In this chapter we will see that γ-ray and x-ray observations
allow us now to exclude the possibility that this new matter is created
either in the form of neutrons or in the form of equal numbers of particles
and antiparticles.

In conclusion, it is fair to say that experimental evidence in recent years
has been in favor of the big-bang theory. Experimental cosmology, however,
is still a very new field which, with the unique tools of space astronomy,
will produce in the coming years some very impressive accomplishments.

8.2 Gamma-Ray Astronomy

Gamma-rays (γ-rays) are photons of extremely high energy. They are
more energetic than x-rays and their domain starts somewhere between
0.1 and 1 MeV. The rest mass of the electron (0.51 MeV) is often taken
as the separation line between x-rays and γ-rays. A head-on collision of
two γ-rays of this energy can produce an electron-positron pair. The domain
of space γ-ray astronomy is approximately the range between 1 MeV and
1 GeV.

Gamma-rays of higher energies can be detected from the ground through
the particle showers they produce in the atmosphere. In the 10^{15} to 10^{20} eV
range these showers are recorded by huge arrays of scintillation detectors
the largest of which is the one built by McCusker in Australia which has
a total collecting area of 250 km. In the 10^{11} to 10^{15} eV range the air
showers are not intense enough to reach the ground but they can be detected
from the visible Cerénkov radiation they produce in the atmosphere.
A 10-meter optical reflector, in the form of a fully steerable dish made up
of 248 hexagonal mirrors, has been installed by the Smithsonian Observatory
in Arizona specifically for this purpose.

We will now proceed to review the different mechanisms for the production and the absorption of γ-rays. The study of these mechanisms will
allow us to appreciate the significance of γ-ray observations in many areas
of astrophysics and cosmology. The most important of the production
mechanisms are:

I *The decay of neutral pions* (π^0)   Neutral pions with a half-life of approximately 10^{-16} seconds decay into two γ-rays. The energy spectrum of these γ-rays peaks around 70 MeV (one half of the 135 MeV rest mass of the π^0 meson) but it extends also over a wide energy range. This process remains actually an effective producer of γ-rays in the entire range from 1 to 10^5 MeV. To trace the astrophysical importance of these γ-rays we must look into the origin in the π^0 mesons. Neutral pions are produced mostly in $p - p$ nuclear interactions, i.e., essentially in collisions of cosmic ray protons with protons (hydrogen) of the interstellar gas.They are also produced in collisions of cosmic ray protons with thermal (optical and infrared) photons in a process called *photopion production*, and appear also as a by-product of *nucleon-antinucleon annihilation* events. These last two processes, however, are not as important as the proton-proton interaction.

If we knew the cosmic ray spectrum and the concentration of neutral hydrogen in the different regions of our galaxy, we could compute the γ-ray flux we would expect to observe in the direction, e.g., of the galactic center, the halo, the arms, etc. At the present time the galactic distribution of atomic hydrogen is rather well known from radioastronomical observations in the 21-cm line, but we have no information about the concentration of molecular hydrogen. The cosmic ray spectrum, on the other hand, is fairly well known only in the vicinity of the earth. It is therefore difficult to predict with any high degree of confidence the expected γ-ray fluxes in the galaxy. In a reverse process, however, if we had good γ-ray measurements toward the different regions of the Milky Way, we could use these observations to improve our knowledge on the concentration of hydrogen and the intensity of the cosmic ray spectrum in our galaxy. A first step in this direction were the observations of Clark, Garmire and Kraushaar (1968) which showed an enhanced γ-ray flux from the disk of our galaxy where there is a higher concentration of neutral hydrogen.

II *Bremsstrahlung*   Relativistic electrons or positrons passing near the nucleus of an atom (much closer than the radial distance of the orbital electrons of the atom) will be decelerated by the Coulomb interaction of the two charged particles and the energy they will lose will be emitted in the form of a photon. For very close encounters, the energy of the photon is roughly equal to the energy of the relativistic electron and therefore electrons with energies greater than 1 MeV are likely to produce γ-rays. This process could be responsible for a significant portion of the galactic γ-ray flux and should be included in all such computations.

III *The inverse Compton effect*   Highly energetic electrons can collide with low energy photons and by transferring part of their energy to the photons they can escalate them to the γ-ray level. If E_T is the total energy of a relativistic electron and m_0c^2 its rest mass, the energy of the photon before ($h\nu_0$) and after ($h\nu$) the collision are related by the expression,

$$h\nu = h\nu_0 \left(\frac{E_T}{m_0c^2} \right)^2 \qquad (8.2\text{-}1)$$

Thus 5 GeV electrons colliding with starlight photons ($h\nu_0 \sim 1$ eV) will produce γ-rays in the 100 MeV range. This is a very important process in many problems of high energy astrophysics and a substantial component of the γ-ray flux we receive might be produced by this mechanism.

Since the energy spectrum of these γ-rays depends among other things on the spectrum of the low energy photons, we might be able to obtain through this process valuable information on the 3°K cosmic background of microwave radiation. Actually people have tried to correlate in this manner reports of a higher than expected γ-ray background and a still unconfirmed excess in the far infrared of the 3°K cosmic radiation. It is also possible that the γ-ray maximum observed toward the galactic center by Clark, Gamire, and Kraushaar (1968) simply reflects through the inverse Compton process, the existence in the galactic center of the intense infrared radiation which has been observed by Low (1970).

IV *Synchrotron Radiation* To produce γ-rays with this process we need strong magnetic fields and extremely relativistic particles. For this reason the synchrotron process can at best be only a minor contributor in the low energy range of the galactic γ-ray spectrum. It can, however, be important in certain x-ray sources where we have strong evidence for the existence of intense magnetic fields and energetic particles.

V *Electron-positron annihilation* This process usually follows the reaction,

$$e^+ + e^- = \gamma + \gamma \qquad (8.2\text{-}2)$$

and produces a γ-ray spectrum with a peak around 0.51 MeV which is the rest mass of the electron and the positron. The detection, or the setting of an upper limit for this line is of great importance to many astrophysical and cosmological problems. A version of the steady state theory, e.g., postulates that particles and antiparticles are continuously produced in equal numbers. If this were true and particles and antiparticles are not kept apart by some unknown mechanism, we would expect to observe a high γ-ray flux from the annihilation of protons with antiprotons, and electrons with positrons. Observations up to now tend to disprove this theory by placing upper limits which are below the γ-ray fluxes expected by this cosmological model.

VI *Nuclear transitions* Nuclei in interstellar matter can be excited to high energy levels by the cosmic rays. The de-excitation of these nuclei results in the emission of γ-rays with specific energies. Gamma rays are also produced in simple nuclear reactions such as in the formation of a deuterium from a proton and a neutron with the emission of a 2.23 MeV γ-ray.

$$p + n = D + \gamma \qquad (8.2\text{-}3)$$

The radioactive materials produced in a supernova explosion must also be strong emitters of γ-radiation.

In addition to the emission mechanisms discussed above, the absorption processes for γ-rays are also of great astrophysical interest because the intensities and spectra we ultimately observe are always a combination of the different emission and absorption processes. The two most important absorption mechanisms for γ-rays are:

I *The Compton effect* In this process, γ-rays lose their energy by colliding with low energy electrons which pick up the energy lost by the γ-rays. This mechanism predominates at energies below 10 MeV but its effectiveness decreases rapidly at higher energies.

II *Pair production* Here γ-rays spend their energy to produce electron-positron pairs. To conserve energy and momentum the production can take place only when the γ-rays hit a particle or another photon. In a head-on collision of a high energy ($h\nu$) photon with a low energy ($h\nu_0$) photon, the electron and the positron produced fly in the direction the γ-ray was moving, each carrying away a total energy E_T where,

$$E_T = \{(m_0 c^2)^2 + (pc)^2\}^{1/2} \qquad (8.2\text{-}4)$$

The conservation of energy requires that,

$$h\nu + h\nu_0 = 2E_T \qquad (8.2\text{-}5)$$

and the conservation of momentum that,

$$\frac{h\nu}{c} - \frac{h\nu_0}{c} = 2p \qquad (8.2\text{-}6)$$

By introducing E_T from (8.2-4) into (8.2-5) and by multiplying (8.2-6) by c we can obtain two equations which when squared and subtracted yield the relation,

$$(h\nu)(h\nu_0) = (m_0 c^2)^2 \qquad (8.2\text{-}7)$$

Equation (8.2-7) gives the threshold for this process, i.e. the lowest energy γ-ray which can produce an electron-positron pair in collision with a thermal photon. But $h\nu_0$ in (8.2-7) is usually of the order of 1 eV or less (the energy of photons from typical thermal sources) and therefore this absorption process becomes important only for γ-rays above 10^{11}–10^{12} eV. At energies above 10^{14} eV, however, this mechanism becomes very effective and the low energy photons ($h\nu_0 \sim 10^{-3}$ eV) of the 3°K cosmic radiation will probably absorb practically all of these very high energy γ-rays.

Gamma-ray detectors such as scintillation counters and Cerenkov counters have been flown up to now on several balloon, rocket and satellite experiments. The much heavier spark chamber counters, which can provide a higher degree of directivity, were carried only on balloons during the 60's because they were too heavy for the available scientific satellites. The weight problem precluded also the use on satellites of more elaborate arrangements of anticoincidence counters or the use of large aperture detectors. Passive collimators, like a honeycomb attachment in front of the detector, were also excluded because the absorption length for even the lower energy γ-rays is several centimeters of lead. As a result, during the 60's all the rocket and satellite borne γ-ray detectors were characterized by a very low angular resolution.

In addition γ-ray astronomy has to cope with extremely low γ-ray fluxes (about one γ-ray photon per hour) and very high background interference levels. These spurious counts are produced on the one hand by the cosmic rays, which can trigger too the γ-ray counters, and on the other hand by the secondary γ-rays which are produced in the terrestrial atmosphere by the cosmic rays. The flux of the secondary γ-rays decreases with altitude, as the particle density of the atmosphere decreases, but even at a depth of only 5 gr/cm^2, i.e. above 99.5% of the atmosphere, it still exceeds the galactic γ-ray flux by a factor of 10. As a result, balloon flights which were able to carry γ-ray detectors of higher directivity had to fight a much higher background noise which essentially nullified the advantage of the larger detectors. It is not surprising, therefore, that the scientific output of γ-ray astronomy during the 60's was rather poor and its results were much less impressive than the results of x-ray astronomy during the same period.

In spite of all these difficulties, however, γ-ray astronomy managed to make considerable progress in the 60's mostly by placing upper limits to different potential γ-ray sources, such as the Crab Nebula, Cygnus A, etc., and by obtaining the first crude measurements of the diffuse galactic γ-ray emission. Placing upper limits can be very important because as we have seen it allows us occacionally to eliminate certain theories. Announcements that a discrete γ-ray source had finally been discovered were made several times during the 60's but these reports were never confirmed by later observations. It is possible that the initial claims were based on a rather hasty inspection of not well documented data. But because there were several such cases and always there was an interval of several months between successive observations, the explanation might be that γ-ray sources are either very variable in time or that they have a very short life span. These ideas are supported by the fact that many x-ray sources have been found to display similar patterns. Gamma-ray observations from satellites in the early 70's will be able to answer these questions through continuous patrol of the γ-ray sky over long periods of time.

Some of the most important γ-ray observations during the 60's were made by Clark, Garmire, and Kraushaar (1968) who flew a rather simple scintillation detector on the OSO-3 satellite. The angular response of this counter from the one half-power point to the other was only 30°, but still it was enough to show, after a very elaborate analysis of the data, that there is an enhanced flux of γ-rays along the galactic disc and especially toward the galactic center. Clark, Garmire, and Kraushaar were also able to obtain an upper limit for the apparently isotropic background flux of γ-rays, in the range above 70 MeV.

In summary, we can say that, in spite of overwhelming difficulties, γ-ray astronomy is finally ready for bigger achievements. In the 70's special satellites, such as the ones in the *Small Astronomical Satellite* (SAS) series will carry into orbit the first spark chambers (Figure 8.2-I) which will improve tremendously the angular resolution of γ-ray observations. These

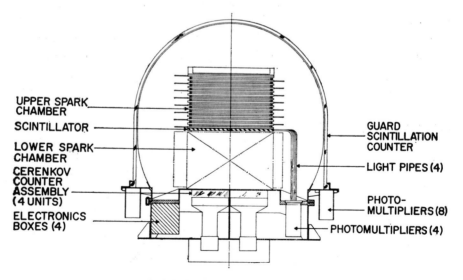

SAS-B GAMMA RAY EXPERIMENT

FIGURE 8.2-I   A spark chamber type gamma-ray detector which can provide
a substantial degree of angular resolution

detectors will have sufficient angular resolution for the proper mapping of the diffuse γ-ray component and hopefully for the detection of several discrete γ-ray sources. Also a Cerenkov telescope with a sensitivity of 10^{-8} photons cm^{-2} sec^{-1} at energies above 500 MeV, and an angular resolution of 1° has been proposed for the study of discrete γ-ray sources in the 70's.

Gamma-rays have the unique advantage over cosmic rays that they are not deflected by magnetic fields, and over other forms of radiation that they can penetrate great depths of intergalactic matter. As a result they can bring direct information from very large distances, i.e., from much older epochs of the universe when matter, at least according to the big-bang theory, was packed much closer together. We have already seen the many areas of astrophysics and cosmology (Fazio, 1967) where γ-ray observations can make an important contribution, and the sophisticated γ-ray observations of the 70's will bring undoubtedly many new, exciting, and as is often the case, totally unexpected discoveries in the years ahead.

8.3 X-Ray Astronomy

X-rays were discovered in 1895 by Roentgen who received the first Nobel Prize in Physics in 1901 for this important discovery. The x-ray spectrum covers primarily the region from 1 to 100 keV, i.e. from about 10 Å to about 0.1 Å, though certain authors extend it up to 0.1 keV on the one side and up to 1 MeV on the other. Hard x-rays ($E > 25$ keV) can penetrate to balloon heights, i.e. down to the 25 to 30 km range. X-rays with energies around 10 keV are stopped near 60 km and the very soft x-rays ($E < 1$ keV) can be observed only above the E-region of the ionosphere.

The detection of the first x-ray source beyond our solar system was achieved accidently in the summer of 1962. Giacconi, Gursky, Paolini, and Rossi (1962), launched an Aerobee rocket with three large Geiger counters hoping to detect fluorescent x-rays from the moon's surface produced by the impact of solar x-rays. The instruments were designed to operate in the 2 to 8 Å range and were at least, 100 times more sensitive than any others flown before. The experiment did not observe any x-rays from the moon, but to everybody's surprise, it detected a strong source of x-rays in the constellation of Scorpius. This first x-ray source was named *Sco XR-1* and its discovery marked the beginning of an exciting new field of astronomy.

By 1971 about 50 discrete x-ray source has been detected, though many of them are extremely variable and a few of them have appeared only for brief periods of time and then have faded away. The Cen XR-2 source is a typical example. After observations in October 1965 had failed to detect this source, in April 1967 it became the brightest x-ray source in the 1–10 keV range. In the following months the brightness of Cen XR-2 decreased steadily and six months later it was no longer detectable. More recently, in July 1969 two Vela satellites detected an x-ray source (Cen XR-4) which had a lifetime of about one month but in this brief interval it managed to surpass by a factor of 2 the brightness of Sco XR-1 in the 3–12 keV range. A more comprehensive study of the variability of the x-ray sources will become possible in the coming years with long-term x-ray observations from satellites. As a minor detail it should be noted that some groups prefer to use only X instead of XR in designating x-ray sources (e.g., Sco X-1 instead of Sco XR-1).

Figure 8.3-1 shows the location of the better known x-ray sources in galactic coordinates. It is easily seen that most of the sources are distributed along the galactic plane, a clear indication that the majority of these sources are members of our galaxy. They also seem to cluster in galactic longitude in two main groups. One is the Sagittarius-Scorpius region toward the galactic center ($l_{II} = 0$) and one in the Cygnus-Cassiopeia region ($l_{II} = 60$ to 120). These groupings suggest that the x-ray sources, like most of the stars, are located predominantly in the different spiral arms of our galaxy and therefore they must also be at a mean distance of not more than 150 par-

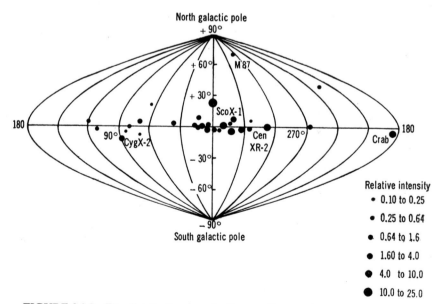

FIGURE 8.3-I   The distribution in galactic coordinates of most of the discrete x-ray sources known up to the end of 1970 (Doyle, 1969)

secs from the galactic equator. On the basis of this argument, from the galactic latitudes we can deduce their mean distance, which turns out to be 2 kpc. Finally from their mean distance we can determine their average luminosity, which is found to be 10^{36}–10^{37} erg/sec in the 1 to 10 Å range. Since an area of 2 kpc in radius is only about 2% of the total area of the galactic disc, our galaxy must have a total of 1000 to 2000 sources of the above average intensity and therefore the total luminosity of our galaxy in the 1–10 Å range must be of the order of 10^{39}–10^{40} erg/sec. For comparison, the luminosity of our galaxy is estimated to be about 10^{38} erg/sec in the radio domain and about 10^{44} erg/sec in the visible region of the spectrum.

Practically all of the 50 or so known x-ray sources were discovered from a few balloon flights and a total of only a few hours observing time with rockets. Gas-filled Geiger and proportional counters, scintillation detectors, and photomultipliers covered with CsI or KCl were the commonly used x-ray detectors. Directivity was obtained either through honeycomb and slot collimators or through the modulation collimator developed by Prof. M. Oda.

The Oda collimator (Figure 8.3-II) consists of two parallel sets of wires with the spaces between the wires approximately equal to the width of the wires. As the collimator rotates, it produces a modulated output signal because a parallel beam of x-rays finds the opening to the detector continuously changing between a maximum and a minimum. The maximum

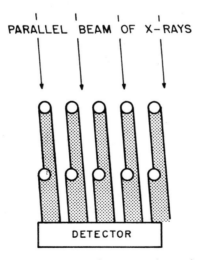

FIGURE 8.3-II   The Oda collimator for x-ray observations of high angular resolution

opening occurs when the open spaces of the front set are in the same line of sight with the open space of the back set, and the minimum occurs when the ray paths through the front set hit the wires of the back set. Modulation collimators can achieve an angular resolution of approximately one minute of arc. They also can provide information about the size of the source because the modulations will decrease as the apparent size of the source increases. The reason, of course, is that the front wires will not cast as sharp a shadow for an extended source as they do with the parallel rays from a point source.

X-rays can be focussed also with special telescopes in which the x-rays are specularly reflected from mirror surfaces at grazing incidence angles. In Section 7.3 we have already described this type of x-ray telescope. Instead, however, of using total external reflection, as in these telescopes, it is possible also to focus x-rays using symmetric Bragg diffraction of x-rays from certain appropriately orientated crystals. Boeing Laboratories have developed such an x-ray telescope by covering the inner surfaces of a series of concentric paraboloidal annuli with thousands of thin, specially machined crystals of lithium fluoride (LiF). It is expected that the Boeing telescope will be operational in the early 70's.

It is expected that several x-ray telescopes, 0,5–1.0 meter in diameter and 5 to 10 meters in focal length, will be placed in orbit on special satellites in the mid and late 70's. It has also been proposed to construct a 5 to 10 m aperture and 50 to 100 m focal length x-ray telescope which will become part of the *National Astronomical Space Observatory* (NASO) in the 80's. There is also discussion of the deployment in space of an array of proportional counters covering an area of up to 100 square feet. This array will be able to map the entire sky in a few days with a resolution of a fraction of a degree and sufficient sensitivity to detect sources 1,000 times weaker than the Crab Nebula.

Up to 1971 only about 10 of the 50 or so known x-ray sources had been identified with optical or radio sources. The reason is that in the uncertainty boxes which contained the x-ray sources there are still too many optical objects. From the few that have been identified with optical objects we have learned that x-ray sources can be of many different types. They can be, e.g., supernovae remnants such as Tau XR-1 in the Crab Nebula, or they can be star-like objects, such as Sco XR-1, which could be an old nova. They can also be entire galaxies, such as Virgo XR-1, where presumably a recent explosion has taken place. Because of the extremely interesting types of astronomical objects that are associated with the x-ray sources, we will proceed to examine the three typical representatives mentioned above in greater detail.

The identification of Tau XR-1 was achieved through a very elegant experiment by Dr. Friedman's group from the Naval Research Laboratory. This group managed to observe the brief lunar occultation of the Crab Nebula on July 7, 1964, with a very accurately timed rocket flight. This was a precious opportunity because these occultations occur only once every nine years. The experiment was a great success and not only did it fix the position of Tau XR-1 with high accuracy, but it also showed that the x-rays were emitted from an extended source in the central region of the visible nebula approximately 1 minute of arc in size.

In 1968 a pulsar was discovered in the Crab Nebula, and soon after that it was shown that simultaneously with the radio pulses it emits optical and x-ray pulses. In the 1–10 keV range, the x-ray pulses represent only about 10% of the total x-ray flux from the Crab Nebula, but the contribution from the pulses seems to increase considerably toward the higher energies (>100 keV) of the x-ray spectrum. The radio, the optical, and the x-ray flux from the entire Crab Nebula are of the same order of magnitude ($\sim 10^{37}$ erg/sec) but the power of the x-ray pulses exceeds the power of the optical pulses by 2 orders of magnitude and the power of the radio pulses by 4 orders of magnitude. This means that in the radio domain the contribution from the pulsar is totally negligible compared to the radio flux from the entire nebula.

The radio and optical emission from the Crab Nebula are undoubtedly due to synchrotron radiation because both of them are strongly polarized. The x-ray emission could also be due to highly energetic electrons, but as seen from (3.4-3), it would require relativistic electrons with energies up to 10^{14} eV to produce x-ray synchrotron emission in the magnetic fields (10^{-3}–10^{-4} Gauss) of the Crab Nebula. This hypothesis cannot be confirmed until larger instruments on orbiting satellites will be able to measure the polarization of the x-ray flux.

As seen from (3.4-6) the half-life time of the highly energetic electrons is quite short and for $E = 10^{14}$ eV it will be of the order of only 3–4 years. But even for the optical synchrotron emission, which has been confirmed and requires relativistic electrons with energies up to 10^{12} eV, the half-life time is only 100 to 200 years. This is still considerably shorter than the nine-century life span of the Crab Nebula and indicates that there must be a continuous supply of energetic particles.

In 1969 it was discovered that the pulse period of the Crab Nebula pulsar increases by about 36.5 nanoseconds per day or about 4.2×10^{-13} seconds per second. Meanwhile a pulsar model had emerged consisting of a rapidly rotating neutron star with a very strong magnetic field of 10^{10} to 10^{12} Gauss. The density of a neutron star is approximately 10^{15} gr/cm^3

and therefore a neutron star of one solar mass will have a radius R of about 7 km. Hence the rotational energy E_R of the Crab Nebula pulsar is,

$$E_R = \frac{2}{5}\frac{1}{2}\left(\frac{2\pi R}{\tau}\right)^2 M \simeq \frac{8MR^2}{\tau^2} \simeq 8 \times 10^{48} \text{ ergs} \qquad (8.3\text{-}1)$$

where in (8.3-1) we have used $M = 2 \times 10^{33}$ gr (one solar mass), $R = 7 \times 10^5$ cm, and $\tau = 0.033$ sec (the measured period of the pulsar).

The decrease in the rotational energy of the neutron star due to the above-mentioned $(d\tau/dt = 4.2 \times 10^{-13}$ sec/sec) slowing down of the period is,

$$\frac{dE_R}{dt} = 16\frac{MR^2}{\tau^3}\frac{d\tau}{dt} = \frac{2E_R}{\tau}\frac{d\tau}{dt} = 2 \times 10^{38} \text{ erg/sec} \qquad (8.3\text{-}2)$$

This energy must be passed continuously from the neutron star to the surrounding nebula through a mechanism which we do not understand very well yet. This amount of energy can easily account for the total emission from the Crab Nebula including the continuous acceleration of about 10^{36} particles/sec to energies of 10^{14} eV $= 160$ erg for the x-ray emission. Over a year, which is the lifetime of these particles, we will have a total of about 3×10^{43} particles, and since as seen from equation (3.4-5) each one of these 10^{14} eV particles will emit $\sim 6 \times 10^{-7}$ erg/sec, the total power emitted would be $\sim 2 \times 10^{37}$ erg/sec which is approximately the total x-ray flux of the Crab Nebula.

It is important to add here that pulsars might turn out to be the main source of practically all the cosmic rays observed. The acceleration of charged particles to such extremely high energies is probably accomplished with the same, not-well-understood mechanism with which the neutron star passes it rotational energy to the surrounding nebula in the form of highly energetic particles.

A charged particle co-rotating with the magnetosphere of a neutron star will reach a velocity very close to the speed of light and therefore super-relativistic energies at a distance,

$$r_c = \frac{c\tau}{2\pi} \qquad (8.3\text{-}3)$$

which is often called the *radius of the light cylinder*. This is also the largest gyroradius R_H that a super relativistic particle spiraling in the magnetic field of the neutron star can attain. We thus have,

$$R_H = \left(\frac{mc}{eH}\right)V = \frac{mc^2}{eH} = \frac{E_T}{eH} = \frac{c\tau}{2\pi} = r_c \qquad (8.3\text{-}4)$$

from which we can find E_T, i.e., the energy limit which can be reached by charged particles in the vicinity of a neutron star. The result is,

$$E_T = 1.4 \times 10^{12} H\tau \qquad (8.3\text{-}5)$$

where E_T is in electron volts, H in Gauss, and τ in seconds. Thus in the early stages of a neutron star's life, when H is probably equal to 10^{10}–10^{12} Gauss and τ is of the order of 10^{-3}–10^{-4} sec, pulsars might be able to produce cosmic rays with energies as high as 10^{18}–10^{20} eV. This would account for almost the entire energy spectrum of the cosmic rays detected in the vicinity of the earth.

Sco XR-1 was the first x-ray source to be discovered (Giacconi *et al.*, 1962). It is the strongest x-ray source in the sky, being approximately 7 times brighter than Tau XR-1 in the 1 to 10 Å range. In 1966 using the Oda collimator a group of scientists from American Science and Engineering and MIT managed to narrow down the location of the Sco XR-1 source into an area of less than a thousandth of a square degree. They also showed that the Scorpius source was less than 20 arc seconds in diameter which, from the extrapolation of the spectrum, meant that it must also be visible as a bluish star of approximately the 13th magnitude. With this information at hand, astronomers of the Tokyo Observatory immediately found the x-ray star and within a week the identification was confirmed by the Palomar Observatory. Optical observations have placed Sco XR-1 at a distance of about 300 pc.

This unusual star emits more than 94 % of its energy in the x-ray region while its radio flux, which is equal to 0.02 *flux units* (1 f.u. = 10^{-26} Watt m^{-2} Hz^{-1}), is barely detectable. Such an emission spectrum could be produced by a hot plasma, with a temperature $T \simeq 5 \times 10^{7}$°K, radiating through thermal bremsstrahlung. Since this is a thermal process the observed frequency spectrum $I(f)$ will be the black-body spectrum $B_f(T)$ modified by the opacity τ of the medium, i.e.,

$$I(f) = B_f(T)(1 - e^{-\tau}) \propto \frac{f^3}{\exp(hf/kT) - 1} \left\{ 1 - \exp\left(\frac{-AN^2}{f^2 T^{3/2}} \right) \right\} \quad (8.3\text{-}6)$$

where, as seen from (2.4-23), the opacity of a fully ionized plasma is proportional to $N^2 f^{-2} T^{-3/2}$. In the x-ray region we usually have $hf/kT \gg 1$ which implies that,

$$\frac{1}{\exp(hf/kT) - 1} = \frac{\exp(-hf/kT)}{1 - \exp(-hf/kT)} \simeq \frac{\exp(-hf/kT)}{hf/kT} \quad (8.3\text{-}7)$$

16*

and therefore the x-ray spectrum will be given by the expression,

$$I(f) \propto f^2(1 - e^{-\tau}) \exp\left(-hf/kT\right) \qquad (8.3\text{-}8)$$

If the electron density N is high, then we are dealing with an *optically thick* medium where $\tau \gg 1$ and $e^{-\tau} \simeq 0$. For this model the spectrum will be,

$$I(f) \propto f^2 \exp\left(-hf/kT\right) \qquad (8.3\text{-}9)$$

If, on the other hand, the electron density is very low, we have an *optically thin* medium where $\tau \ll 1$ and $(1 - e^{-\tau}) \simeq \tau$. In this model, since τ is proportional to f^{-2}, the spectrum will be,

$$I(f) \propto \exp\left(-hf/kT\right) \qquad (8.3\text{-}10)$$

The x-ray data we have on Sco XR-1 are not yet precise enough to allow us to clearly choose one of these two models. Data seem to favor a compact ($\sim 10^9$ cm), high density (10^{16} cm^{-3}) plasma, though a huge volume of very tenuous plasma is still a possibility also.

Studies of old Harvard plates have indicated that the brightness of the star of the Scorpius source has remained at the same general level for the past 70 years. If this is the case with the x-ray source also, then here again, as in the case of the Crab Nebula, we must have a continuous replenishment of energy by some still unknown central source. Sco XR-1, however, is an irregular variable over its entire spectrum. Its weak radio flux can vary by a factor of 20 in only one to two hours, its V and B magnitudes can changes by 0.3 magnitudes per hour in the 12.2 to 13.2 magnitude range, and its x-ray flux can vary by a factor of 4 in less than an hour. It has been observed that the visible star becomes bluer when the x-ray intensity increases. In some ways the Sco XR-1 star resembles an old nova, which as in the case of many other novae might be a close binary with a stream of gas flowing from one star to the other.

In one of the models that have been proposed, one star of the binary is a neutron star which, with its enormous gravitational field, accelerates the accreted gas to energies capable of x-ray emission. In this model the side of the companion star facing the neutron star must have a temperature close to 30,000°K, due to the high x-ray flux it receives. This would explain the blue color of the visible star and the fact that it becomes bluer when the x-ray flux increases. Unfortunately, neither the optical nor the x-ray flux of Sco XR-1 show any periodic variation as one would expect from a binary. It is possible, however, that we do not observe these variations because the orbital plane of the binary is nearly perpendicular to the line of sight. It should be noted that Cyg XR-2, which is about 20 times weaker than Sco XR-1 but resembles it in many ways, has been tentatively identified with a very short period (about 6 hours) spectroscopic binary. Still, however, there are many more crucial observations to be made and much more to be learned from this class of x-ray sources.

Very few x-ray sources have yet been identified with extragalactic objects. The clearest case is Virgo XR-1 which has been identified with the galaxy M 87, which is also a powerful radio source (Virgo-A). M 87 is a giant elliptical galaxy with probably 10 times or more stars than our own galaxy. A long jet streaming out of M 87 is a clear indication that this galaxy has undergone a recent violent explosion. Cygnus-A and Centaurus-A, which, like Virgo-A, are strong radio galaxies and also bear signs of past explosions, have been searched for x-ray emission. The results for Cygnus-A have been negative, and Centaurus-A has an x-ray emission 10–20 times weaker than Virgo-A.

Since the radio flux of Virgo-A and Centaurus-A, as measured from the earth, are of the same order of magnitude and the radio flux of Cygnus-A is ten times larger, it follows that not all radio galaxies are strong x-ray sources. The answer might be the time that has elapsed from the galactic explosion which produced the radio sources. In most radio galaxies the radio emission comes from two regions on either side of the optical galaxy, and it is assumed that the matter in these regions was ejected from the galaxy at the time of the explosion. Thus, by measuring the distance of the two radio sources from the optical center of the galaxy we can speculate about the age of these sources. The available data indicate that Virgo-A is the youngest while Cygnus-A is by far the oldest. It is possible, therefore, that the x-ray emission occurs only for a relatively short period following a galactic explosion.

The quasar 3C273 has been identified also as an x-ray source. If the distance estimated from the red shift of this quasar is correct ($\sim 2 \times 10^9$ l.y.), then its x-ray emission is of the order of 10^{46} erg/sec, i.e., more than a million times the estimated x-ray emission of our galaxy. Still the contribution from these few strong x-ray galaxies seems to be considerably less than the total x-ray emission from all the normal galaxies like ours. A Seyfert galaxy (NGC 1275), which is also the Perseus-A (3C84) radio source, was identified in 1971 as one more extragalactic x-ray source. This is at a distance of about 200 million l.y. and is estimated to emit also about 10^{46} erg/sec in the x-ray region.

An area of great interest is also the diffuse and isotropic x-ray background that has already been observed. Undoubtedly part of it is due to weak, unresolved sources. It has been estimated, however, that if on the average all galaxies were like our galaxy, the total contribution would be less than 10% of the observed isotropic background, though admittedly this percentage could be enlarged by including evolutionary effects.

Another possible explanation is that the diffuse x-ray flux in the 1 keV to 1 MeV range is produced by cosmic ray electrons acting on the photons of the 3°K cosmic radiation through the inverse Compton effect. At energies lower than 1 keV there is evidence (Friedman, 1969) that the x-ray background is due to thermal radiation (bremsstrahlung) from a hot ($\sim 10^{6}$°K), rarefied ($\sim 10^{-5}$ protons/cm^3) plasma that seems to fill the intergalactic space. If this interpretation is correct, then we must be living in a closed universe because the resulting average density of 2×10^{-29} gr/cm^3 is sufficient to reverse the present expansion of the universe some time in the future. The fact that x-ray observations might be able to detect the missing matter of the universe gives them a tremendous astronomical importance.

The diffuse x-ray flux in the 1 keV to 1 MeV range can be described by an exponential law,

$$N \text{ (photons cm}^{-2} \text{ sec}^{-1} \text{ sr}^{-1} \text{ MeV}^{-1}) \propto E(\text{MeV})^{-\alpha} \quad (8.3\text{-}11)$$

where $\alpha \simeq 2$, though the spectral index might be somewhat different in the different energy domains. An extrapolation of the above spectrum seems to fit also with the γ-ray background observed by the OSO-3 near 100 MeV. The x-ray background in the 2 to 8 Å range, is approximately equal to 10 photons cm^{-2} sec^{-1} sr^{-1}. One version of the steady state theory claims that new matter is continuously created in the form of neutrons. The electrons, however, produced from the β-decay of these neutrons would heat up the intergalactic medium to a temperature of about 10^{9}°K and would make it emit a thermal x-ray flux five times higher than the one we observe. Thus, our measurements of the diffuse x-ray background allow us to eliminate one more cosmological model. A more precise study of the diffuse x-ray and γ-ray spectrum (Fazio, 1970) is of high importance because it contains cosmological information of great value.

8.4 Ultraviolet Space Astronomy

The energy required to ionize a neutral hydrogen atom is 13.6 eV which corresponds to a wavelength of 912 Å. Ultraviolet radiation in the Lyman continuum ($\lambda < 912$ Å) can ionize the hydrogen atoms and therefore it is strongly absorbed by the neutral hydrogen which fills the interstellar space. As a result the spectral domain of galactic space ultraviolet astronomy extends from about 3,000 Å, where the ozone absorption of the terrestrial atmosphere starts, to 912 Å, where the absorption by the interstellar hydrogen begins. Solar ultraviolet observations do not have this lower limit because the total amount of neutral hydrogen between the earth and the sun is too insignificant to cause any noticeable absorption.

Instruments for space ultraviolet observations must overcome many technical difficulties. Thus, lenses made out of fused silica become opaque below 1,800 Å. Other materials can be used at shorter wavelengths but the final limit is reached with lithium fluoride (LiF), which becomes opaque at 1,050 Å. Reflecting telescopes can work in principle at even shorter wavelengths, but the reflecting efficiency of most commonly used reflectors decreases rapidly with wavelengths in the 1,500 to 1,000 Å range. As a result it was necessary to develop mirrors and grating spectrometers from new materials, such as platinum, beryllium, etc.

Detectors used in UV observations include: ionization chambers with different gases, photographic plates with special emulsions, and photo-multiplier tubes with a LiF window or of the open type for work at $\lambda < 1,050$ Å. Image storage tubes sensitive in the ultraviolet region (Uvicon tubes) have also been used on the *Orbiting Astronomical Observatory* (OAO-2). These tubes combine essentially the advantages of photographic plates and photo-multipliers because they can store electronically on a very fine grid a whole picture which can be recorded and transmitted like a television picture.

The first crude survey of the sky in the ultraviolet was attempted with a 20° beam by an NRL group in 1955. The early observations were made from spinning rockets which made practically impossible the detailed study of any particular source. Later on, however, when stabilized rockets with a high pointing accuracy became available, observers were able to obtain the ultraviolet spectra of many individual stars and nebulae with spectral resolutions better than 1 Å (Smith, 1969). Satellite observations of stellar UV radiation did not begin until the mid-sixties. Ultraviolet astronomy has placed a lot of its expectations in the *Orbiting Astronomical Observatories* (OAO). Unfortunately, OAO-1, which was launched in 1966, did not obtain any data due to an early failure in the power supply. OAO-3, which was launched at the end of 1970 carrying a 36-inch telescope, failed to reach a proper orbit and was destroyed before returning any data. These two very costly failures (hundreds of millions of dollars) have been strong setbacks for space astronomy.

These disappointments have been ameliorated by the almost flawless performance of OAO-2, which was launched in December of 1968. OAO-2 was probably the largest (~2 tons), most expensive, and most sophisticated scientific satellite launched in the 60's. It was a harbinger of the complex space observatories planned for the 70's, and from its nearly circular 770 km orbit has sent back an unprecedented wealth of ultraviolet observations. The OAO-2, which is shown diagrammatically in Figure 8.4-I, had an octagonal frame approximately 10 feet long and 7 feet in diameter.

All the astronomical instruments were located inside a central cylinder, 40 inches in diameter, which ran the entire length of the frame. The power for all the experiments was provided by the solar panels which measured 21 feet from tip to tip. The OAO-2 was able to point toward any position on the sky with an accuracy of one minute of arc, and then by locking onto a certain number of guiding stars and using its flywheels, it was able to maintain this orientation with a precision of one arc second. The OAO-2 carried two ultraviolet experiments, one by the University of Wisconsin (Code, 1969) and the other by the Smithsonian Astrophysical Observatory (Davis, 1968). The two observing systems occupied the two opposite ends of the central cylinder mentioned above.

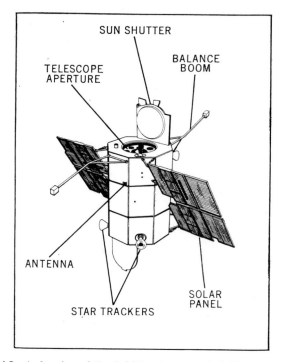

FIGURE 8.4-I   A drawing of the Orbiting Astronomical Observatory (OAO-II) satellite (Papagiannis, 1969)

The University of Wisconsin payload consisted of seven telescopes, the largest of which was a 16-inch reflector. The spectral resolution of the Wisconsin experiment was no better than about 10 Å, which is not suitable for fine spectral work. It is sufficient, however, for observing several strong absorption lines such as the resonance line of triply ionized silicon (Si IV) near 1,400 Å, and the resonance line of triply ionized carbon (C IV) near

1,550 Å. Also easily detectable near 1,200 Å is the Lyman-α line of neutral hydrogen, produced by the interstellar hydrogen.

The Smithsonian payload, which was called the *Celescope* project, consisted of four telescopes, all 12.5-inches in diameter and 24-inches long. Ultraviolet radiation from a sky region 2.5 degrees in diameter was focussed by each telescope on an image storage tube. The trade name for these tubes was *Uvicon* to denote that they had a surface sensitive to the UV spectrum. The spectral regions of the four telescopes were: 3,200–2,200 Å, 3,200 to 1,600 Å, 2,000–1,350 Å, 2,000–1,050 Å. The objective of the experiment was to measure the far ultraviolet brightness of many thousands of stars and thus compile the first ultraviolet star catalogue of the skies. These observations, together with the corresponding optical data, will be used in long range statistical studies of stellar evolution and other important astrophysical problems. A preliminary analysis of the data has shown that diffuse nebulae, which are extremely large gas clouds where new stars are formed, radiate much stronger in the ultraviolet than what the young stars forming within them could account for. The reason might be that at wavelengths shorter than 2,000 Å the dust particles, which are very abundant in these clouds, stop absorbing and start reflecting the electromagnetic radiation falling upon them.

One of the main objectives of ultraviolet astronomy is the study of stars of the O, B, and A spectral classes. These stars are considerably hotter than our sun and emit most of their energy in the ultraviolet region. The hotter stars are of particular interest because they evolve much faster. Furthermore, the intense ultraviolet radiation which they emit ionizes the interstellar gas to very large distances (*H II regions*), this changing significantly the physical state and the dynamical properties of the interstellar medium.

OAO-2 data, based on the strength of the Si IV and C IV lines, indicate that stars earlier (hotter) than B 1 must have temperatures higher than what is presently accepted. Hence a new adjustment of the temperature scale of the very hot stars must be made, starting with a temperature of approximately 28,500°K for a B 0.5 normal dwarf or main sequence star.

Silicon stars, on the other hand, have been found by the OAO-2 to be considerably fainter in the ultraviolet comparing to other stars of the same spectral class. This deficiency is probably due to the additional opacity in the ultraviolet produced by ionized silicon and magnesium in the atmospheres of these stars.

In addition to the early type (hot) stars discussed above, ultraviolet astronomy has an interest also in the later type stars because, as we know from our sun, we can expect to find coronal and chromospheric emission lines in their ultraviolet spectra. This might also be the method to study flare activity in other stars. Recent OAO-2 observations have shown that late-type (red) supergiants are considerably brighter than expected in the ultraviolet. This is attributed now to the chromospheric emission lines, primarily Mg II and Fe II, which when not resolved contribute significantly to the total ultraviolet emission of the cooler stars.

One of the most impressive discoveries of ultraviolet astronomy has been that certain very hot supergiants are ejecting a continuous stellar wind with speeds of the order of 1,000 to 1,500 km/sec, which exceeds by far the escape velocity of these stars. This conclusion was based on the Doppler shifts of absorption lines, such as Si IV and C IV, that are observed in the ultraviolet spectra of such stars. The line profiles of these absorption lines indicate that huge quantities of matter are blown continuously into the interstellar space. It is estimated that the loss rate of mass in these stars can be as high as 10^{-5} solar masses per year. This must be not only an important factor in the evolution of these massive hot stars, but also a significant factor in the dynamics and the replenishment of the interstellar medium.

Ultraviolet observations of high spectral resolution can provide extremely valuable information about interstellar molecules. Typical in this respect is the first detection in 1970 of interstellar molecular hydrogen (H_2) by G. Carruthers of the Naval Research Laboratory. This discovery was made in ultraviolet observations toward Xi Persei and a preliminary analysis of the data suggests that at least in this direction the densities of atomic and molecular hydrogen are comparable.

Ultraviolet observations can also answer many questions about the nature of interstellar dust. These small particles reflect, scatter, and absorb the light from the stars causing interstellar extinction and interstellar reddening. From the spectral dependence of interstellar extinction we can derive useful information about the sizes of the dust particles and about their chemical and physical composition. Present ultraviolet observations show that the interstellar extinction peaks near 2,200 Å but the intensity and the width of the peak seem to vary in the different directions of the galaxy, possibly because the nature of the dust particles is not uniform over the entire galaxy.

One of the theories that have been proposed, suggests that the dust particles are actually small grains of graphite covered with a coating of solid molecular hydrogen. Other theories suggest that they might be a mixture of C_2, SiC, and $MgSiO_3$ particles. More spectral studies and polarization observations in the coming years will help clear these questions. It is very important to understand the composition and density of interstellar matter not only because it has been ejected from certain types of stars and will be used to form others, but also because the correct interpretation of stellar spectra and stellar emission depends on the proper evaluation of the effects imposed by the interstellar medium.

The objectives of space ultraviolet astronomy, besides the study of common stars and galaxies, include also the investigation of a large variety of other, relatively rare celestial objects. The importance of these objects in our studies of stellar and galactic evolution outweighs by far the fact that they are found only in very small numbers. These not very common sources, some of which are very strong emitters of UV radiation, are:

1   Early type stars with extended atmospheres such as P Cygni and Wolf-Rayet stars.

2   Planetary nebulae, which are extensive, often expanding envelopes around some intensely hot stars.

3   Cepheid variable stars, including the RR-Lyrae type with periods of less than one day. Also different types of flare stars.

4   Magnetic stars such as the peculiar A stars.

5   Eclipsing binaries such as the 31 Cygni with one of the two stars an early type dwarf and the other a late type giant. We can study the physics and chemistry of the extensive atmosphere of the cooler giant star by observing the ultraviolet light from the hot dwarf star as it passes behind successive layers of the giant atmosphere.

6   Old novae and supernovae, especially those that are also x-ray emitters, such as Sco XR-1 and the Crab Nebula, where in the latter the pulsar provides an additional strong point of interest.

7   Different types of abnormal galaxies such as the M 82 and M 87 which show signs of recent violent explosions, the Seyfert galaxies with their bright, starlike nuclei, and the puzzling quasars that are believed to be the furthest objects we have detected in the universe.

Finally it is important to search for the unexpected also. It is advisable, for example, to look into the UV spectra of sources without any spectral lines because the reason we have not discovered any blue-shifted objects up to now might simply be that all the lines of these objects are shifted into the ultraviolet which is not accessible to ground observations.

8.5   Optical and Infrared Space Astronomy

The optical window of the terrestrial atmosphere extends from about 3,000 Å, where ozone makes the atmosphere opaque, to approximately 10,000 Å = 1 μ (micron), where the water vapor absorption starts becoming a problem. Actually the atmosphere remains fairly transparent up to 1.3 μ but beyond this wavelength it becomes opaque except for a few relatively free windows (1.5 to 1.75 μ, 2,0 to 2.4 μ, 3.4 to 4.2 μ, 4.5 to 5.0 μ, 8 to 13 μ, and 17 to 22 μ) between the different absorption bands of H_2O and CO_2.

The effectiveness of these windows for ground observations varies considerably with time and improves decisively with altitude because their width and transparency depend primarily on the water vapor content of the terrestrial atmosphere. From the last infrared window to the beginning of the radio spectrum, i.e. from about 25 μ to about 1,000 μ = 1 mm, the atmosphere is totally opaque. The infrared domain extends from 1 to 1,000 μ and is usually divided into the near (1–4 μ), the intermediate (4–25 μ) and the far (25–1,000 μ) infrared.

In the optical region, where the atmosphere is essentially 100% transparent, ground-based telescopes are beset by the brightness of the sky and the turbulance of the atmosphere. In addition of course they have to cope with the weather and the problems of city lights and air contamination. In an ideal optical system the first dark ring around the image of a point source has an angular diameter 2α (in radians) given by the expression,

$$2\alpha = 1.22 \frac{\lambda}{\beta} \qquad (8.5\text{-}1)$$

where λ is the observing wavelength and β the radius of the lens or the mirror. Approximately 84% of the light collected by the instrument falls inside this first dark ring and therefore the image of a point source appears to extend over an area $\pi\alpha^2$ on the celestial sphere. For a 120-inch (3-meter) telescope and λ = 5,000 Å, α is equal to 0,04 second of arc and the area a point source will appear to occupy on the celestial sphere is equal to 0.005 square arc second. The turbulance of the atmosphere, however, makes it impossible to obtain values of α any better than 0.5 arc second and disperses the image of a point source over an area of almost a whole square second of arc. The result is that a mere 10-inch telescope in space will be able to match the angular resolution of the 200-inch telescope of Mt. Palomar.

The airglow of the night sky is equivalent to a star of the 22nd magnitude per square arc second. Since the atmosphere spreads the image of any point source over almost a whole square second of arc, we cannot expect to detect sources much dimmer than of the 23rd magnitude. Space telescopes will be able to detect much fainter sources. The diffuse background brightness of the sky above the terrestrial atmosphere is like a star of the 23rd magnitude per square arc second, which brings a gain of one stellar magnitude. The biggest gain, however, comes from the fact that space telescopes will be able to focus the image of a star into a much smaller area and thus reduce decisively the amount of background emission competing with the light of the star. The 120-inch *Large Space Telescope* (LST), which is now under study and hopefully in the late 70's or early 80's will become part of a permanent space observatory, will be able to detect stars of the 29th magnitude. Note that a difference of 5 stellar magnitudes is equal to a factor of 100 in intensity which is essentially gained by focussing the image of a star into an area of 0.005 square arc sec. which is about 100 times smaller than the focusing area of the ground telescopes. The LST will improve also the resolving power of all ground-based telescopes by at least a factor of 10. Finally it is of interest to mention that large space telescopes will be much easier to construct mechanically because the weightlessness of space will eliminate the extremely elaborate structural supports that are necessary on the ground.

Early attempts to place telescopes above the major part of the earth's atmosphere used balloons and in Section 7.4 we have seen the successful results of Stratoscope-I. Much more impressive, however, were the observations of Stratoscope-II which was flown by the same Princeton group in 1963, 1964 and 1970. A balloon with a volume of about 5 million cubic feet was used in these flights to lift a 36-inch telescope to altitudes of 85,000 ft. ($\sim$26 km) in order to extend our astronomical observations into the infrared region. Stratoscope-II had excellent infrared optics and was an outstanding example of how a space telescope can be remotely controlled from the ground to perform a large variety of astronomical observations without planning everything in advance. During its second flight in November 1963, Stratoscope-II obtained excellent spectra in the 1 to 3 μ range of eight red giant stars and in some of them such as in Betelgeuse (α-Orionis), it detected for the first time strong bands of water vapor (Woolf, Schwarzschild and Rose, 1964). In the 1970 flight it made some very intriguing observations of the Seyfert galaxy NGC 4151 which is at a distance of approximately 30 million light years.

Infrared telescopes are now flown also on special jet airplanes which can reach altitudes in the 40,000 to 50,000 ft. range. At these heights, conditions are far superior to even the best ground sites and have made infrared observations possible even in the 25–100 μ range. Infrared detectors have also been flown on rockets and were used on all Mariner missions to Mars and Venus. Infrared astronomy will undoubtedly be performed ultimately from large satellites, but for the time being there is still a lot to be gained from ground observation through the different atmospheric windows and from the relatively inexpensive and very practical (easy retrieval of data and instruments) jet flights. As a matter of fact a 36-inch telescope is being prepared for use on one of these jet planes and a lot of ground-based telescopes are now being outfitted for infrared observations.

Rayleigh scattering in the atmosphere decreases rapidly with wavelength (actually this is the reason why the sky looks blue) and at $2\,\mu$ it is already 100 times weaker than at $0.5\,\mu$. As a result ground observations through the 2.0–$2.4\,\mu$ window can be conducted even during the day. Infrared observations, however, especially at the longer wavelengths are confronted with a different, much more difficult problem. The problem is that everything, including the entire sky and the telescope itself, glows with infrared radiation. Bodies at room temperature ($\sim 300°K$) radiate with a maximum intensity around $10\,\mu$ ($T\lambda_{max} \simeq 0.3$). As a result, trying to perform infrared observations at $10\,\mu$ is like trying to make optical observations during the day with a luminescent telescope.

The radiation from the instruments can be reduced by cooling them with dry ice, liquid nitrogen, liquid hydrogen, or even liquid helium. The effects of the background emission on the other hand can be greatly reduced by setting the primary or the secondary mirror of the telescope into a periodic motion (rocking) while keeping the detector fixed. As the mirror rocks about its central position, the source under observation moves in and out of the field of the detector producing an alternating current. The background emission, on the other hand, that reaches the detector is not affected by the small motions of the mirror and therefore it generates a continuous (direct) current. By ignoring the d.c. component and amplifying only the a.c. component of the detector current we can detect quite faint discrete sources against a bright diffuse background.

The first systematic mapping of the sky in the infrared was undertaken in 1967 by a group from the California Institute of Technology under Neugebauer and Leighton (1968) using the 2.2 window of the atmosphere. This study has shown that of the 5,500 brighter stars catalogued in the infrared only 30% had a counterpart in the nearly 6,000 stars visible to the human eye. The correspondence dropped to only 1% when the 300 faintest objects in the two groups were compared. Obviously these results mean that there is a great wealth of new information in the infrared domain and one wonders why this region of the spectrum had been neglected for such a long time by ground observers. One of the reasons, of course, was the erroneous belief that all sources are essentially thermal and therefore there is nothing much to be gained from observations in other regions of the spectrum. The main problem, however, was that we did not have sensitive enough infrared detectors and spectrometers.

Modern infrared detectors are essentially either photodetectors or thermal detectors, both cooled to very low temperatures. Photodetectors use the solid-state equivalent of the photoelectric effect and usually have a limited spectral range of operation. A typical example is PbS which, refrigerated to liquid nitrogen temperatures (77°K), is commonly used in the 1 to 4 μ range. Thermal detectors, on the other hand, have electric resistances that change rapidly with temperature and by converting the infrared radiation into heat they measure minute changes in temperature. The spectral sensitivity of the thermal detectors covers practically the entire infrared spectrum. A typical example is a germanium crystal dopped with impurities, such as gallium or copper, which when refrigerated by liquid helium to a temperature of about 2°K is capable of measuring temperature changes as small as 10^{-7}°K. The recent development of the Josephson detectors (Ulrich, 1969), which are junctions of two superconductors connected by a weak link, is a great step foward for infrared observations in the very interesting submillimeter region.

Infrared spectroscopy made also tremendous progress during the 60's. Scanning the diffraction spectrum from a grating with a detector is a very inefficient process because the detector can use only a very small portion of the incoming radiation. Consequently, as we try for higher spectral resolution, especially in the case of weak sources, we reach very rapidly the sensitivity limit of our detectors. Fourier spectroscopy, on the other hand, uses all the incoming radiation and in the recent years has improved astronomical infrared spectroscopy by a factor of 100 making possible spectral resolutions in the infrared of the order of $\Delta\lambda/\lambda = 10^{-4}$.

Fourier spectroscopy was used for the first time at the turn of the century by Michelson with his famous interferometer, but it had to be rediscovered in the 50's when fast electronic computers made its full application possible. In simple terms, in Fourier spectroscopy the light from the source is divided into two beams which follow paths of different length to the detector. As the path difference is varied, light at different wavelengths is enhanced and diminished through constructive and destructive interference. As a result the output of the detector contains all the information of the spectrum of the incoming radiation, which in a sense is scanned as we vary the path difference of the two beams. A Fourier transform of the record obtained by the detector, which nowadays is done with fast electronic computers, reconstructs the spectrum of the analyzed radiation.

The development of telescopes, detectors, spectroscopic devices, cryogenic systems, etc., for infrared astronomy is advancing now at a fast pace. This rapid progress in infrared technology, together with the fact that jet aircrafts, balloons, rockets, and finally satellites have become available for infrared observations, guarantee a very exciting future for infrared astronomy.

Infrared radiation suffers very little extinction by the interstellar dust and therefore it is a unique tool for the study of our galaxy. At 2.2 μ, e.g., we can see red stars up to 2 kpc away and we can observe the strong infrared source at the center of our galaxy which is at a distance of approximately 10 kpc. Most of the very cool stars ($T < 2,000°K$) that have been detected up to now are long period variables with periods ranging from a few months to several years. The infrared brightness of some of these variables has been observed to change by a factor of 10 over a period of three years. There are of course also many pseudo-infrared stars, i.e. stars which actually are much hotter than they appear to be from their red color. The light of these stars, which are not variables, has simply been reddened by the interstellar dust which eliminates the blue component of the starlight and lets through the red.

The most interesting category of infrared sources in our galaxy are stars in the very early stages of their lives. These *protostars* are still surrounded by parts of the dense dust cloud from which they were formed. Theoretical computations show that protostars before settling into their position in the main sequence, they go through a brief period of very high luminosity when their temperatures becomes approximately equal to their ultimate main sequence temperature, while their size is still much larger. Protostars in this stage cannot be seen optically because all of their radiation is absorbed

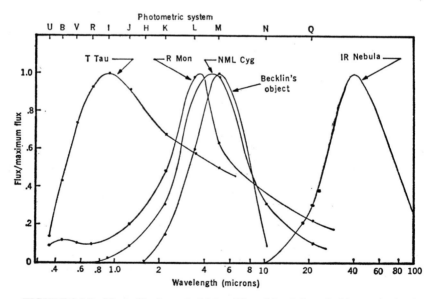

FIGURE 8.5-I   Normalized spectral intensities of five infrared objects, obtained by the University of Arizona (Low, 1969)

by the surrounding dense cloud of dust. They can be observed, however, in the infrared where the dust cloud reemits the energy it absorbs.

Figure 8.5-I shows the spectra of different infrared sources which are believed to belong to the protostar category. The R Monocerotis source seems to be in the preplanetary stage when material, from which the planets are ultimately formed, is still distributed as dust around the bright protostar. Radio observations have detected strong OH (the hydroxyl radical) emission from the NML Cygnus source, which is one of the brightest infrared objects in the sky. The infrared star that was discovered by Becklin in the Orion nebula is also an OH-source. Near Becklin's object, Low and Kleinman have discovered an infrared nebula which is the brightest object in the sky at 22 μ. In spite of its low temperature ($\sim 100°$K), the extent and total luminosity of this infrared nebula are such that the most plausible source of energy is a cluster of protostars. T-Tauri stars, one of which is also shown in Figure 8.5-I, represent the final stage of stellar formation just before these stars join the main sequence.

Planetary nebulae, which are expanding gaseous clouds surrounding a very hot central star, have also been found to be strong infrared sources. Planetary nebulae, like NGC 7027, should not be confused with the protostars which are also surrounded by a gas and dust cloud. Planetary nebulae are produced by a loss of matter from stars in a late stage of their evolution.

A strong source of infrared emission has now been located at the dynamical center of our galaxy and its position coincides with a strong radio source (Sagittarius-A) which had previously been identified with the galactic center. The radiation comes from a small volume, only about 1 light year in radius, and peaks near 100 μ in the far infrared. The total power radiated by this source is of the order of 10^{41}–10^{42} erg/sec, equivalent to the energy output of 10^8 stars like our sun. It is interesting to note that in the vicinity of the sun, which is typical of most of the galaxy, a volume one light year in radius contains on the average less than one star. The Andromeda Nebula, our neighboring galaxy which very much resembles our own galaxy, appears to have a very similar infrared emission from its galactic center.

These ordinary galaxies are actually very weak emitters of infrared radiation compared to some other rather unusual galaxies. Unstable galaxies, such as the exploding galaxy M 82, have much stronger infrared sources in their galactic centers. Seyfert galaxies, which are characterized by their extremely bright starlike nuclei, emit in excess of 10^{46} erg/sec in the infrared, with their spectra peaking in the far infrared. Observations with Stratoscope-II in 1970 have shown that the nucleus of the NGC 4151 Seyfert galaxy is less than 12 light years in diameter. Such high luminosities represent an energy output which is 10–100 times higher than the total energy produced by our galaxy. Finally 3 C 273 and other quasars seem to be even stronger emitters of infrared radiation. It was a great surprise to discover that these quasars emit most of their enormous energy in the far infrared.

One can say that there seems to be a sequence from quasars, to Seyfert galaxies, to unstable galaxies, and finally to mature galaxies like our own, in which as the importance of the galactic center with respect to the entire galaxy diminishes, we observe a rapid decrease in the brightness of the source and an enormous reversal in the ratio of the infrared to the optical emission of the galaxy. This, some people believe, might represent an age sequence in the evolution of galaxies.

The mechanism by which such enormous amounts of radiation are produced in such small volumes, as the quasars and galactic centers, is still very difficult to understand. It is also difficult to see why most of the radiation is emitted in the far infrared and why the intensity of all these sources displays always considerable time fluctuations. Frequent collisions among millions or even billions of stars packed in a volume 1–10 light years in diameter has been suggested as one possible explanation. It has been suggested also that the energy might be coming from a high rate of supernova explosions in these young and dense galactic nuclei, or it might be the energy generated by a large number of pulsars which were produced in these explosions. Some people have discussed even more exotic processes, such as the continuous creation and annihilation of matter and antimatter at the centers of the galaxies (Low, 1970).

There is no doubt that as space infrared astronomy advances to maturity we will learn a great deal more about some very fundamental problems, such as the formation of stars and planetary systems, and the mysterious processes going on in the galactic centers. Also observations of the infrared background in the submillimeter region, i.e., in the 100 to 1,000 μ range, will help us resolve the problem of where the so-called 3°K background radiation actually peaks and therefore if it really is a 3°K radiation.

8.6   Space Radio Astronomy

Radio astronomical observations at frequencies below the critical frequency of the ionosphere (the highest plasma frequency of the ionospheric layers) cannot be made from the ground because the ionosphere reflects these waves back into space. Even at twice the critical frequency, however, the strong diffraction and absorption of the ionosphere and intense scintillation effects

caused by the ionospheric irregularities, render ground observations difficult and often impractical. One should not forget also that the penetration
frequency is inversely proportional to the cosine of the zenith angle
($f_m = f_c/\cos \chi$) and thus for a source at 60° from the zenith it will be twice
the critical frequency of the ionosphere. The cut-off frequency of the ionosphere for vertical incidence varies from about 1 MHz to about 10 MHz
depending on the time of the day, the season, the phase of the solar cycle
and the latitude of the station. Ground observations have been pushed
on a few occasions to frequencies close to 1 MHz, but these observations
were beset by large uncertainties and were only night-time observations
in the direction of the local zenith. In general ground radio astronomy is
limited to frequencies above 20 MHz.

Radio astronomical observations in the 0.1 to 10 MHz range must be
carried out above the maximum of the ionosphere, which occurs at around
300 km, and therefore they require either satellites or high altitude rocket
probes. The instrumentation is relatively simple consisting of a very sensitive radio receiver and a dipole antenna, typically 10 to 100 ft. long. Theoretical studies and actual observations have shown that the behavior of
an antenna in a plasma is considerably different from what it is in free
space. The plasma effects become stronger as the observing frequency
approaches the local plasma frequency and they are very complex when,
as in the case of the terrestrial ionosphere, there is an external magnetic
field present (magnetoactive plasma). Theoreticians, aided by experimental
results, have tried to predict the behavior of the antenna impedance under
such critical conditions but the problem is very difficult and has not yet
been solved in its entirety. For this reason experimenters prefer to conduct
their observations at the highest possible altitude in order to approach
free space conditions as much as possible. At 1,000 km the plasma frequency
and the gyro frequency are both of the order of 1 MHz, while at a distance
of one earth radius (6,400 km) the plasma frequency is of the order of
250 kHz and the gyrofrequency is of the order of 150 kHz. Consequently
observations at frequencies below 300 kHz, i.e., at wavelengths longer than
1 km, must be conducted beyond the plasmapause, i.e., at a distance of
several earth radii even at high latitudes. The observing frequencies can
not be extended below a certain limit because even the interplanetary
space has a plasma density of 5–10 el/cm³. This corresponds to a plasma
frequency (2.3-12) of about 25 kHz which represents the final limit of
space radio observations.

Space radio astronomy is also faced with a technical problem which is
very difficult to overcome. This is the very low directivity of all antennas
used for space radio observations. The beam width θ of an antenna (in

17*

radians) is given by the expression,

$$\theta \simeq \frac{\lambda}{L} \qquad (8.6\text{-}1)$$

where λ is the observing wavelength and L the length of the antenna. When L is shorter than a wavelength, the antenna becomes almost omnidirectional and is not capable of resolving the radiation coming from the different regions of the sky.

In the hectometer range ($100 < \lambda < 1{,}000$ m) a typical dipole antenna on a rocket probe or on a satellite is almost always much smaller than the wavelength, and therefore most observers (e.g., Huguenin and Papagiannis, 1965) were able to measure only the integrated (average) radio background of the entire sky.

The first major step in gaining higher directivity for space observations was made with the launching of Explorer 38, the first *Radio Astronomy Explorer* (RAE-1) satellite in 1968. RAE-1 carried two oppositely directed V-type antennas, each leg of the V being 229 meters long. The two V-antennas, which formed an X with a 60° acute angle, provided a gravity gradient stabilization for the satellite which as a result had always one V-antenna pointing toward the earth and the other toward the local zenith of the celestial sphere. RAE-1 was equipped also with a 37-m dipole antenna which bisected the obtuse angle of the X and therefore was always in a horizontal position.

The 400 lb. satellite was placed on a nearly 6,000 km circular retrograde orbit at a 59° inclination to the equator. This type of orbit allowed RAE-1 to remain for 6 months of the year in sunlight thus avoiding the thermal stresses on the long antennas from passing continually in and out of the earth's shadow. A television camera was carried on board to monitor the distortions of the long V-antennas due to the differential gravity forces and to their uneven heating by the solar rays. The sunlit sides of the long antenna rods become hotter, and as they expand more than the dark sides, they produce a curving of the long rods. Due to the long computations required and the limited computer time available, the reduction of the data from the long V-antennas has been slow and difficult. Fortunately, however, the shorter dipole antenna has provided some very fine data (Alexander *et al.*, 1969) and it is hoped that in due time some good results will be retrieved also from the records of the V-antennas.

A second RAE satellite with similar antennas is scheduled for the early 70's and might be placed in an orbit around the moon. Plans exist also for even larger antenna systems that can provide a major increase in directivity at frequencies around 1 MHz. Typical of these ambitious projects is the huge rhombic antenna (Figure 8.6-I) which has been studied by the University of Michigan under a grant from NASA. The diameter of this antenna is 10 km, but the whole structure will weigh less than 1 ton. The rhombic antenna forms two unidirectional and independent beams, 180° apart, directed along the long axis of the rhombic. At 1 MHz each beam will focus on an area of about 80 square degrees, which is less than 0.2% of the entire celestial space.

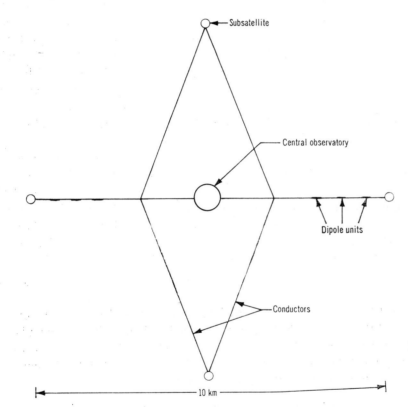

FIGURE 8.6-I   The large rhombic antenna designed by the University of Michigan for space radio astronomical observations of high directivity in the hectometer wavelength range (Doyle, 1969)

The diffuse radio emission of our galaxy is synchrotron radiation produced by energetic electrons of primary and secondary cosmic ray origin spiralling in the weak magnetic fields (10^{-5}–10^{-6} Gauss) of the galaxy. Cosmic ray observations in the vicinity of the earth have shown that the energy spectrum of the cosmic ray electrons is given by the power law,

$$N(E)\, dE = KE^{-\gamma}\, dE \qquad (8.6\text{-}2)$$

where $N(E)$ is the number of electrons with energies between E and $E + dE$, and K and γ are constants. Theoretical computations show that the frequency spectrum of the synchrotron radiation emitted by these particles in a magnetic field H is given by the expression,

$$I(f) = Cf^{\left(\frac{1-\gamma}{2}\right)} H^{\left(\frac{1+\gamma}{2}\right)}$$

(8.6-3)

where C is a constant and γ is the exponent of (8.6-2). At high energies γ has been measured to be close to 2.5, but at lower energies, where direct measurements are difficult to perform, it seems that $N(E)$ begins to depart from the power law of (8.6-2). The energy content of the cosmic ray gas depends critically on its low energy component which can best be evaluated by the low-frequency synchrotron radiation it emits. Measurements, therefore, of the galactic radio spectrum at low frequencies are of great importance in understanding the cosmic ray spectrum with all its cosmological implications.

The radio spectrum we observe includes also absorption effects and therefore it is not exactly the same as the emission spectrum of (8.6-3). The most important factor is *free-free absorption* in which a photon is absorbed by an electron to allow the electron to increase its velocity as it swings-by near a proton. Free-free absorption occurs when the radiation passes through an H II region, i.e., a region of ionized hydrogen. In Section 2.4 we have seen that the opacity τ of an ionized region is proportional to f^{-2},

$$\tau = Af^{-2}$$

(8.6-4)

where A is a constant which depends on the electron density, the temperature and the path length through the ionized region. If the absorbing layer is located somewhere between the source and the observer, as seen from (A-38), the brightness spectrum received $I_b(f)$ is related to the emitted spectrum $I'(f)$ of (8.6-3) through the expression,

$$I_b(f) = I'(f) e^{-\tau} = CH^{(1+\gamma)/2} f^{(1-\gamma)/2} \exp(-Af^{-2})$$

(8.6-5)

As the frequency decreases the absorption increases exponentially while the emitted spectrum increases only through a power law. Consequently, after reaching a maximum while the opacity is still not very high, $I(f)$ will start decreasing exponentially with the frequency. By differentiating (8.6-5) with respect to f and equating the derivative to zero, we can find the frequency f_{max} at which $I(f)$ becomes maximum. The result is,

$$f_{max} = \left(\frac{4A}{\gamma - 1}\right)^{1/2}$$

(8.6-6)

Space observations (Figure 8.6-II) have shown that the maximum of the galactic radio spectrum occurs in the range between 2 and 5 MHz. The upper and lower curves of Figure 8.6-II bracket the range which includes most of the observations reported in the literature. The width of this range reflects, on the one hand the error bars of the available data, and on the other hand the fact that, in spite of their low directivity, different observations were directed toward a different general region of the galaxy.

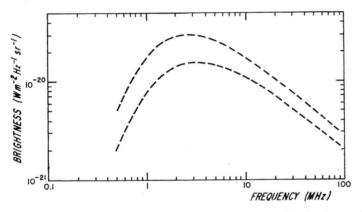

FIGURE 8.6-II   The brightness spectrum of the cosmic radio background in the long radio wavelength region. The upper and lower curves bracket the experimental uncertainties of the observations as well as the differences in the spectra of the different regions of the Milky Way

If, instead of an absorption layer, we are actually dealing with a uniform mixture of emitting and absorbing regions, as seen from (A-43), the spectrum received is given by the expression,

$$I_b(f) \simeq I''(f)\left(\frac{1 - e^{-\tau}}{\tau}\right) \qquad (8.6\text{-}7)$$

where $I''(f)$ is the non-thermal emission (8.6-3), and in (8.6-7) we have neglected the much weaker thermal emission. The maximum can be found again by differentiating this equation, but the important difference is that at the low frequencies, where τ becomes large and $e^{-\tau}$ tends to zero, the frequency dependence of $I_b(f)$ will be,

$$I_b(f) \simeq \frac{I''(f)}{\tau} = \frac{C_0 f^{(1-\gamma)/2}}{A f^{-2}} \propto f^{(5-\gamma)/2} \qquad (8.6\text{-}8)$$

Equation (8.6-8) shows that a uniform mixture will have a slower (power law) decrease than the steep (exponential) decrease of an absorbing layer. It should be mentioned that synchrotron self-absorption (Ginzburg and Syrovsatski, 1969) is also a possible absorption mechanism. In this case the spectrum decreases as $f^{-5/2}$ which, since $\gamma \simeq 2.5$, is considerably steeper than the f^{-1} to $f^{-3/2}$ of (8.6-8).

At high frequencies ($f > 100$ MHz) the radio sky appears to be dark as a whole with a bright belt of radio emission along the galactic disc and an even brighter region toward the galactic center. On the contrary, at low frequencies ($f < 10$ MHz) the radio sky will appear to be very bright with a dark band, due to high absorption along the galactic plane and an even darker area toward the center of the galaxy. One can almost say that the high and low frequency maps of the radio sky will be like the positive and the negative copies of a photograph. Space radio astronomical observations might provide the means to distinguish between the halo and the extragalactic contribution to the overall sky brightness. This will be a very important contribution because it is still unclear whether the Milky Way has a halo or not.

As space antennas gain in directivity first we will be able to study separately the different regions of the galaxy such as the galactic center, disc, and halo, and later to study individually different discrete galactic and extragalactic sources of great interest such as pulsars, radiogalaxies and quasars. Coherent emission phenomena occur only at wavelengths that are longer than the mean distance of the plasma particles. It is therefore quite possible that at these long wavelengths we will be able to detect coherent emission from certain astrophysical plasmas, which are undoubtedly present in many of these very intriguing radio sources. In conclusion we must add that besides the orbiting of large antenna structures in space it is also possible that we will be able to acheive high angular resolution at these long wavelengths by using two independent satellites as a radio interferometer. One can also foresee the deployment of a long antenna arrays on the moon. These arrays will probably be located on the back side of the moon to avoid the radio noise from the earth and especially the potential gyro-radiation in the hectometer range from the Van Allen belts.

Some of these future plans sound almost like dreams, or science fiction, but most probably they are only 10 to 20 years away. Space astronomy offers the unique opportunity to study the astronomical universe from wavelengths longer than 10^5 cm to wavelengths shorter than 10^{-13} cm. There is no doubt that as the necessary technology develops, space astronomy will lead us to many astounding discoveries and will generously reward the quest of man to understand the universe in which he dwells.

8.7 Bibliography

A. Books for Further Study

1. *Astrophyiscs and Stellar Astronomy*, T. L. Swihart, John Wiley & Sons, New York, N. Y., 1968.
2. *Exploration of the Universe*, G. Abel, Holt, Rinehart and Winston, New York, N. Y., 1969.
3. *A Long-Range Program in Space Astronomy*, ed. by R. O. Doyle, NASA SP-213, U. S. Govt. Printing Office, Washington, D. C., 1969.
4. *Space Astronomy* 1958–1964, NASA SP-91, U. S. Govt. Printing Office, Washington, D. C., 1966.
5. *Manned Laboratories in Space*, ed. by S. F. Singer, Springer-Verlag, New York, N. Y., 1969.
6. *Non-Solar X- and Gamma Ray Astronomy*, ed. by L. Gratton, IAU Symposium No.37, Springer-Verlag, New York, N. Y., 1970.
7. *Astronomy in Space*, H. Newell, H. Smith, N. Roman and C. Mueller, NASA SP-127, U. S. Gov. Printing Office, Washington, D. C., 1967.
8. *Space Research: Directions for the Future*, Space Science Board, NAS/NRC, Publ. No. 1403, Washington, D. C., 1966.
9. *Electromagnetic Radiation in Space*, ed. by J. G. Emming, Springer-Verlag, New York, N. Y., 1967.
10. *Introduction to Space Science*, ed. by W. N. Hess and G. D. Mead, Gordon and Breach, New York, N. Y., 1968.
11. *High Energy Astrophysics*, T. C. Weeks, Chapman and Hall, London, England, 1969.
12. *Telescopes in Space*, Z. Kopal, Faber and Faber, London, England, 1968.
13. *Les Observatoires Spatiaux*, J.-C. Pecker, Presses Universitaires de France, Paris, France, 1969.
14. *Infrared Astronomy*, ed. by P. J. Brancazio and A. G. W. Cameron, Gordon and Breach, New York, N. Y., 1968.

B. Articles in Scientific Journals

Alexander, J. K., L. W. Brown, T. A. Clark, R. G. Stone, and R. R. Weber, The spectrum of cosmic radio background between 0.4 and 6.5 MHz, *Astrophys. J.*, **157**, L 163, 1969.

Clark, G. W., G. P. Garmire and W. L. Kraushaar, Observation of high-energy cosmic gamma rays, *Astrophys. J.*, **153**, L 203, 1968.

Code, A. D., Photoelectric photometry from a space vehicle, *Publ. Astron. Soc. Pac.*, **81**, 475, 1969.

Davis, R. J., The Celescope experiment, Spec. Rep. 282, Smithsonian Astrophys. Obs., Cambridge, Mass., 1967.

Fazio, G. G., Gamma radiation from celestial objects, *Ann. Rev. Astron. Astrophys.*, **5**, 481, 1967.

Fazio, G. G., High-energy gamma ray astronomy, *Nature*, **225**, 905, 1970.

Feldman, P. A., M. J. Rees and M. W. Werner, Infrared and microwave astronomy, *Nature*, **224**, 752, 1969.

Friedman, H., Cosmic x-ray observations, *Proc. Roy. Soc.*, London, A313, 301, 1969.

Giacconi, R., H. Gursky, F. Paolini and B. Rossi, Evidence for x-rays from sources outside the solar system, *Phys. Rev. Letters*, **9**, 439, 1962.

Ginzburg, V. L., and S. I. Syrovatski, Developments in the theory of synchrotron radiation and its reabsorption, *Ann. Rev. Astron. Astrophys.*, **7**, 375, 1969.

Huguenin, G. R., and M. D. Papagiannis, Spaceborne observations of radio noise from 0.7 to 7.0 MHz and their dependence on the terrestrial environment, *Annales d'Astrophys.*, **28**, 239, 1965.

Low, F. J., Infrared astrophysics, *Science*, **164**, 501, 1969.

Low, F. J., The infrared phenomenon, *Astrophys. J.*, **159**, L 173, 1970.

Neugebauer, G., and R. B. Leighton, The infrared sky, *Sci. Amer.*, **219**, August, 1968.

Papagiannis, M. D., Space Astronomy, *Science*, **166**, 775, 1969.

Smith, A. M., Rocket spectrographic observations of α-Virginis, *Astrophys. J.*, **156**, 93, 1969.

Spitzer, L., Astronomical research with the large space telescope, *Science*, **161**, 225, 1968.

Ulrich, B. T., Millimeter wavelength observations of solar activity, the moon, and Venus with a Josephson junction detector, *Bulletin, AAS*, **2**, 222, 1970.

Wilson, R., and A. Boksenberg, Ultraviolet astronomy, *Ann. Rev. Astron. Astrophys.*, **7**, 421, 1969.

Woolf, N. J., M. Schwarzschild, and W. K. Rose, Infrared spectra of red-giant stars, *Astrophys. J.*, **140**, 833, 1964.

RADIATIVE TRANSFER
AND THE EDDINGTON APPROXIMATION

The Specific Intensity and the Flux of Radiation

The unit commonly used to measure radiation in astrophysical problems in the *specific intensity* $I_\nu(\theta)$. The definition of this unit is a little complicated and therefore it might be useful to try to relate it to a simple example. Let us imagine that we are watching a game of croquet, where several players are trying to pass balls, each of a certain color, under a series of hoops which have a width dl. One way of measuring how the game is progressing is to count in a given period of time the number of balls of a certain color that make it through a hoop and then follow a generally favorable direction toward the next hoop. So in order to count a ball, not only must it pass through the hoop, but its trajectory must also remain inside a small angle $d\theta$ centered along the most favorable direction F which makes an angle θ with the normal to the crossing line.

Let us now extend this example into three dimensions so that $d\theta$ will become the solid angle $d\Omega$, and dl will become the area element dA (Figure A.1-I). If we now replace the croquet balls of a certain color by photons of a given frequency interval ν to $\nu + d\nu$, we obtain essentially the definition of the intensity of radiation. The only logistic improvement in the actual definition of the intensity $I_\nu(\theta)$ is that instead of counting individual photons, since they all have the same energy $h\nu$, we simply measure the total energy E_ν. Thus the units of the specific intensity are erg cm^{-2} sec^{-1} sr^{-1} Hz^{-1}, and $I_\nu(\theta)$ represents the energy E_ν, in the form of photons in the frequency interval ν to $\nu + d\nu$, which crosses a unit area dA per unit time dt, and remains inside a solid angle $d\Omega$ centered along an axis which makes an angle θ with the normal to the area dA.

If the frequency dependence is not important in a particular problem, we can integrate $I_\nu(\theta)$ over all frequencies and replace it with $I(\theta)$. Finally sometimes we find it useful to eliminate the directional dependence of $I(\theta)$ by averaging over all angles to obtain the *mean intensity J.*

$$J = \frac{1}{4\pi} \iint I(\theta) \, d\Omega = \frac{1}{2} \int I(\theta) \sin\theta \, d\theta \qquad (A-1)$$

Knowing $I(\theta)$ we can find the *flux density* F, i.e., the number of photons, or better, the total energy which crosses *at right angles* a unit area per unit time. From the definition of F it follows that,

$$F = \iint I(\theta) \cos\theta \, d\Omega = 2\pi \int I(\theta) \cos\theta \sin\theta \, d\theta \qquad \text{(A-2)}$$

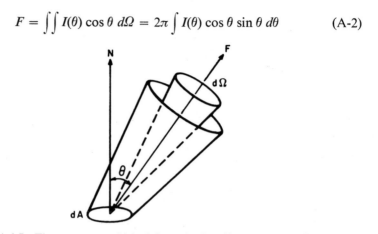

FIGURE A.1-I   The geometry which defines the Specific Intensity $I_\nu(\theta)$

The momentum crossing per unit time a unit area at an angle θ to the normal is equal to the flux in this direction divided by the speed of light c. Hence it is equal to $I(\theta) \cos\theta \, d\Omega/c$, and its component normal to the surface will be $I(\theta) \cos^2\theta \, d\Omega/c$. But the time change of momentum per unit area is pressure, and therefore, integrating over all angles we obtain the *radiation pressure P*,

$$P = \frac{1}{c} \iint I(\theta) \cos^2\theta \, d\Omega = \frac{2\pi}{c} \int I(\theta) \cos^2\theta \sin\theta \, d\theta \qquad \text{(A-3)}$$

In (A-1), (A-2), and (A-3) we have averaged $I(\theta)$, $I(\theta) \cos\theta$, and $I(\theta) \cos^2\theta$ over a whole sphere. When $I(\theta)$ is independent of the angle θ, i.e., for isotropic radiation where $I(\theta) = I$, we simply find that $J = I$, $F = 0$, and $P = \dfrac{4\pi I}{3c}$. This, however, can not be exactly the case in a planetary or stellar atmosphere, because $F = 0$, i.e., zero net flux means that no radiation is advancing toward the top of the atmosphere and therefore no radiation will be emitted from the top of the atmosphere.

The next simplest approximation is to assume that $I(\theta)$ has only two constant values, one for all the upward moving radiation, so that for $0 < \theta < 90°$.

$$I(\theta) = I_u \qquad \text{(A-4)}$$

and one for all the downward moving radiation, so that for $90° < \theta < 180°$,

$$I(\theta) = I_d \qquad \text{(A-5)}$$

This approximation (Figure A.1-III) has been very useful in dealing with problems of radiation in stellar and planetary atmospheres. Using (A-4) and (A-5), in (A-1), (A-2), and (A-3) we get,

$$J = \frac{1}{2}(I_u + I_d)$$ (A-6)

$$F = \pi(I_u - I_d) = F_u - F_d$$ (A-7)

$$P = \frac{2\pi}{3c}(I_u + I_d) = \frac{4\pi}{3c}J$$ (A-8)

The Equation of Radiative Transfer

Let us now consider a beam of radiation transversing a layer of thickness dh at an angle θ to the vertical (Figure A.1-II). The intensity $I_\nu(\theta)$ of the radiation will be reduced by absorption in the layer and will be enhanced by emission from the medium. The amount absorbed per unit volume is proportional to $I_\nu(\theta)$ and to the density ϱ of the medium. The constant of proportionality $\varkappa_\nu$ is called the *mass absorption coefficient*. The amount emitted per unit volume, on the other hand, is equal to the *mass emission coefficient j_ν* times the density ϱ of the medium. Since the distance transversed by the beam inside the layer is $dh/\cos\theta$, the intensity of the radiation after crossing the layer at an angle θ to the normal will change by an amount $dI_\nu(\theta)$ where,

$$dI_\nu(\theta) = -I_\nu(\theta)\varrho\varkappa_\nu \frac{dh}{\cos\theta} + j_\nu\varrho \frac{dh}{\cos\theta}$$ (A-9)

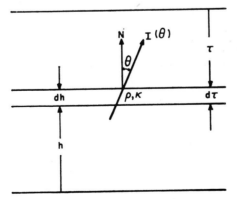

FIGURE A.1-II   A beam of radiation transversing an atmospheric layer of thickness dh at an angle θ to the vertical

It is now convenient to introduce the *opacity* τ_v of the medium, which sometimes is also called the *optical thickness* or the *optical depth* of the medium. The opacity τ_v is defined by the relation,

$$d\tau_v = -\varrho \varkappa_v \, dh \qquad\qquad (A\text{-}10)$$

where the minus sign is used because the optical depth is measured downwards from the top of the atmosphere where $\tau_v = 0$ (Figure A.1-II). Using (A-10) in (A-9) we get,

$$\cos\theta \, \frac{dI_v\,(\theta)}{d\tau_v} = I_v(\theta) - \frac{j_v}{\varkappa_v} \qquad\qquad (A\text{-}11)$$

The atmospheres of the stars and the planets are usually in *local thermodynamic equilibrium* (LTE) which means that the average intensity of radiation at each point of the atmosphere is equal to the emission of a black-body at the local temperature T. Under LTE, $\varkappa_v$ and j_v are related through Kirchhoff's law,

$$\frac{j_v}{\varkappa_v} = B_v(T) \qquad\qquad (A\text{-}12)$$

where $B_v(T)$ is the intensity of the black-body radiation,

$$B_v(T) \, dv = \frac{2h v^3}{c^2} \, \frac{1}{e^{(hv/kT)} - 1} \, dv \qquad\qquad (A\text{-}13)$$

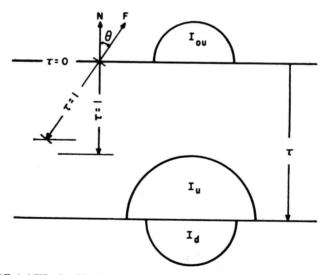

FIGURE A.1-III   In this simplifying approximation the directional dependance of $I(\theta)$ is taken to be $I(\theta) = I_u$ for all upgoing radiation and $I(\theta) = I_d$ for all downgoing radiation

Thus under LTE the equation of radiative transfer becomes,

$$\cos\theta\,\frac{dI_\nu(\theta)}{d\tau_\nu} = I_\nu(\theta) - B_\nu(T) \qquad \text{(A-14)}$$

Note that since black-body radiation is isotropic, $B_\nu(T)$ is not a function of θ as is $I_\nu(\theta)$.

When the absorption coefficient is independent of the wavelength, we say that we have *gray matter*. For a *gray atmosphere* we can integrate (A-14) over all frequencies to obtain,

$$\cos\theta\,\frac{dI(\theta)}{d\tau} = I(\theta) - B(T) \qquad \text{(A-15)}$$

where, as we know from Planck's law,

$$\pi B(T) = \sigma T^4 \qquad \text{(A-16)}$$

Stellar and planetary atmospheres are usually treated as plane, horizontally stratified layers because, as a rule, their thickness is much smaller than the radius of the respective star or planet. Energy moves from the lowest layers to the top of the atmosphere from which it is radiated into the surrounding space. The energy in most cases is transferred from one atmospheric layer to the next through radiation, rather than through conduction or convection. This is called *radiative transfer*. When an atmosphere is in *radiative equilibrium* the flux remains the same at all depths because otherwise we would have accumulation of energy at certain layers. This means that in a horizontally stratified atmosphere,

$$\frac{dF}{d\tau} = 0 \qquad \text{(A-17)}$$

It should be noted here that only the total energy flux remains constant, and radiative equilibrium does not imply that $dF_\nu/d\tau_\nu$ is also equal to zero. The constant flux F, which passes continually through the different layers and is ultimately emitted from the top of the atmosphere, determines the *effective temperature* T_e of the star or the planet. T_e is the temperature at which a black-body would emit the same flux. Hence T_e is defined through the relation,

$$F = \sigma T_e^4 \qquad \text{(A-18)}$$

The Eddington Approximation

We will now try to solve the equation of radiative transfer (A-15) using the approximation of (A-4) and (A-5) and assuming, as Eddington did, that at large optical depths $I_u \simeq I_d$. By integrating (A-15) over all angles and using (A-1) and (A-2) we get,

$$\frac{dF}{d\tau} = 4\pi J - 4\pi B(T) \tag{A-19}$$

Furthermore, assuming that the atmosphere is in radiative equilibrium, where as seen from (A-17) $dF/d\tau = 0$, we find from (A-19) that,

$$J = B(T) = \frac{\sigma}{\pi} T^4 \tag{A-20}$$

Next, we can multiply (A-15) by $\cos \theta$ and then integrate it over all angles. The result, using (A-2) and (A-3), is,

$$c \frac{dP}{d\tau} = F \tag{A-21}$$

because,

$$\iint B(T) \cos \theta \, d\Omega = 2\pi B(T) \int_0^\pi \cos \theta \sin \theta \, d\theta = 0 \tag{A-22}$$

Since, as we have seen, F is constant with respect to τ we can integrate (A-21) to obtain,

$$cP = F\tau + C_0 \tag{A-23}$$

where C_0 is the integration constant. Replacing now P with J from (A-8) we get,

$$\frac{4\pi}{3} J = F\tau + C_0 \tag{A-24}$$

At the very top of the atmosphere (Figure A.1-III) there is no downward moving radiation because above the top of the atmosphere there is only the free space. As a result, from (A-6) and (A-7) we have that at $\tau = 0$, $J_0 = \frac{1}{2} I_{0u}$ and $F_0 = \pi I_{0u}$. Hence setting $\tau = 0$ in (A-24) we get,

$$C_0 = \frac{4\pi}{3} J_0 = \frac{2\pi}{3} I_{0u} = \frac{2}{3} F_0 \tag{A-25}$$

But F is the same at all layers hence F_0 is the same as F and thus (A-24) becomes,

$$\frac{4\pi}{3} J = F\tau + \frac{2}{3} F \tag{A-26}$$

or,

$$J = \frac{F}{\pi} \left(\frac{1}{2} + \frac{3}{4} \tau \right) \tag{A-27}$$

Since at $\tau = 0$ we have $I_d = 0$, and at large optical depths we expect to have nearly isotropic conditions, i.e., at $\tau \gg 1$ we have $I_u \simeq I_d$, it follows from (A-27) and (A-6) that,

$$I_d = \frac{F}{\pi}\left(\frac{3}{4}\tau\right) \qquad\qquad \text{(A-28)}$$

and

$$I_u = \frac{F}{\pi}\left(1 + \frac{3}{4}\tau\right) \qquad\qquad \text{(A-29)}$$

Using now (A-18) and (A-20), we can express (A-27) in terms of T and T_e. The final result is,

$$T = T_e\left(\frac{1}{2} + \frac{3}{4}\tau\right)^{1/4} \qquad\qquad \text{(A-30)}$$

This is the famous *Eddington approximation* which describes the change of the temperature T_e with the optical depth τ. Note that the effective temperature T_e occurs at $\tau = 2/3$ and not at $\tau = 0$. At the top of the atmosphere, i.e. at $\tau = 0$, the temperature T_0 is,

$$T_0 = T_e\left(\frac{1}{2}\right)^{1/4} = 0.86T_e \qquad\qquad \text{(A-31)}$$

Now that we have obtained the relation between T and τ, we can use it to solve the equation of radiative transfer (A-15). The integral solution of this differential equation is,

$$I(\theta, \tau) = e^{\tau \sec \theta} \int_{\tau}^{\infty} B(T)\, e^{-\tau' \sec \theta} \sec \theta \, d\tau' \qquad\qquad \text{(A-32)}$$

which for $\tau = 0$, i.e., for the radiation coming out from the top of the atmosphere, yields the integral,

$$I(\theta, 0) = \int_{0}^{\infty} B(T)\, e^{-\tau' \sec \theta} \sec \theta \, d\tau' \qquad\qquad \text{(A-33)}$$

But from (A-20), (A-30), and (A-18) we have,

$$B(T) = \frac{\sigma}{\pi}\, T^4 = \frac{\sigma}{\pi}\, T_e^4\left(\frac{1}{2} + \frac{3}{4}\tau\right) = \frac{F}{2\pi}\left(1 + \frac{3}{2}\tau\right) \qquad \text{(A-34)}$$

and by introducing (A-34) into (A-33) we can perform the integration to obtain the expression,

$$I(\theta, 0) = \frac{F}{2\pi}\left(1 + \frac{3}{2}\cos \theta\right) \qquad\qquad \text{(A-35)}$$

This final result was derived using the Eddington approximation and assuming local thermodynamic equilibrium, radiative equilibrium, and a horizontally stratified gray atmosphere. The fact that we have used a step-function dependance of I on θ to derive (A-35) should not be considered as a contradiction or as an inconsistency. It is only like using simpler tools to construct others that are more complex and more precise.

From a pragmatic point of view the directional dependence of (A-35) reflects the fact that the radiation emitted in the different directions originates essentially at different depths in the atmosphere. The bulk of the radiation emitted from the top of the atmosphere originates at an optical depth $\tau = 1$ below the top of the atmosphere. As seen from Figure A.1-III, the point which yields an opacity equal to unity occurs closer to the top of the atmosphere for rays propagating at larger angles θ. We have seen, however, that the temperature increases with depth and therefore the rays which propagate closer to the vertical will come from deeper, and hence hotter layers of the atmosphere. Since hotter bodies emit more intense radiation, $I(\theta, 0)$ will have a maximum at $\theta = 0$, which is in agreement with the result we have obtained. The relative decrease of $I(\theta, 0)$ with θ is given by the expression,

$$\frac{I(\theta, 0)}{I(0, 0)} = \frac{2}{5} + \frac{3}{5} \cos \theta = 1 - 0.6 + 0.6 \cos \theta \qquad \text{(A-36)}$$

which, as seen in Section 4.2, is in very good agreement with experimental observations of the limb-darkening effect of the solar photosphere.

Radiative Transfer in Radio Astronomy

When a source of intensity I'_ν is behind an absorbing region of opacity τ_ν, the equation of radiative transfer (A-14) in this region and in the line of sight ($\theta = 0$) becomes,

$$\frac{dI_\nu}{d\tau_\nu} = I_\nu \qquad \text{(A-37)}$$

which has the solution,

$$I_\nu = I'_\nu e^{-\tau_\nu} \qquad \text{(A-38)}$$

In the more general case, in computing I_ν we must include also the emission of the absorbing layer. For some generality we can assume that this region is a uniform mixture of thermal plasma and of non-thermal radio sources. Considering only normal incidence ($\theta = 0$), (A-11) in this case takes the form,

$$\frac{dI_\nu}{d\tau_\nu} = I_\nu - \frac{j_\nu}{\varkappa_\nu} - \frac{j''_\nu}{\varrho \varkappa_\nu} \qquad \text{(A-39)}$$

where $\varkappa_v$ and j_v are the free-free mass absorption and mass emission coefficients of the thermal plasma and j_v'' the volume emission coefficient of the non-thermal radiation. The solution of this differential equation is,

$$I_v = I_v' \, e^{-\tau_v} + \int_0^{\tau_v} \left(\frac{j_v}{\varkappa_v} - \frac{j_v''}{\varrho \varkappa_v} \right) e^{-\tau'_v} \, d\tau'_v \qquad \text{(A-40)}$$

where τ_v is, as we mentioned above, the opacity of this region. In such problems we usually assume a uniform temperature T_0 for the entire plasma region. In this case j_v, j_v'', and $\varkappa_v$ are independent of τ_v and can be taken out of the integral to give,

$$I_v = I_v' \, e^{-\tau_v} + \left(\frac{j_v}{\varkappa_v} - \frac{j_v''}{\varkappa_v} \right) (1 - e^{-\tau_v}) \qquad \text{(A-41)}$$

For a thermal plasma $j_v/\varkappa_v = B_v(T_0)$, and for the non-thermal emission we have,

$$\frac{j_v''}{\varrho \varkappa_v} = \frac{j_v'' L}{\varrho \varkappa_v L} = \frac{I_v''}{\tau_v} \qquad \text{(A-42)}$$

where L is the thickness of this region. Hence (A-41) can be written in the form,

$$I_v = I_v' \, e^{-\tau_v} + \left[B_v(T_0) + \frac{I_v''}{\tau_v} \right] (1 - e^{-\tau_v}) \qquad \text{(A-43)}$$

In radioastronomical problems we have, as a rule, that $h\nu/kT \ll 1$, and therefore (A-13) is simplified to the Rayleigh-Jeans formula,

$$B_v(T_0) \, d\nu = \frac{2kT_0}{\lambda^2} \, d\nu \qquad \text{(A-44)}$$

Using similar formulas to relate I_v, I_v', and I_v'' to their equivalent temperatures T_b, T', and T'', we can convert (A-43) to a relation of temperatures,

$$T_b = T' \, e^{-\tau_v} + \left(T_0 + \frac{T''}{\tau_v} \right) (1 - e^{-\tau_v}) \qquad \text{(A-45)}$$

where T_b is the *brightness temperature* we observe with our antennas. In the simple case of a thermal plasma region (HII-region), without I_v' or I_v'', (A-45) becomes,

$$T_b = T_0 (1 - e^{-\tau_v}) \qquad \text{(A-46)}$$

When τ_v tends to infinity (optically thick medium) we simply have $T_b = T_0$, while when τ_v tends to zero (optically thin medium) T_b also tends to zero. An application of this equation in the case of the solar corona is given at the end of Section 4.3.

THE DEVELOPMENT OF THE SPACE AGE

The beginning of the space age can be traced to the early adventures of the imagination of many scientists and science fiction writers. Johannes Kepler, e.g., in his "Somnium" (Dream) imagined travelling to the moon on the wings of his demons using as a pathway the shadow connecting the earth with the moon during a lunar eclipse. Much more realistic were the space adventures described in the past century by H. G. Wells and especially by Jules Verne. The latter, with his prophetic foresight had even placed the launching facilities for the first lunar expedition in the state of Florida and the splash-down site in the Pacific Ocean.

At the turn of the century, the Russian Konstantin Tsiolkovsky made the first serious calculations of space travel and started developing the theoretical equations of rocket propulsion. During the first third of the 20th century enthusiastic pioneers such as the German Hermann Oberth and the American Robert Goddard constructed and tested the first real rockets with both solid and liquid fuels. In the years before and during World War II the Germans advanced the rocket technology for military purposes and toward the end of the war they were able to bombard London with their V-2 rockets, after long perfecting tests at Peenemunde on the Baltic coast. After the end of the second World War, the United States and the Soviet Union, both drawing heavily on the captured German rockets and the transplanted German scientists and technicians, started a serious, government-sponsored program in rocket development.

Naturally the scope of this program was of a military nature but science started benefiting even from the early test flights by mounting on them modest scientific instruments. The first rocket (a captured V-2) with an american scientific payload was launched on May 10, 1946, from White Sands, New Mexico. It reached a maximum altitude of about 110 km and provided interesting data about the upper atmosphere. The first rocket flight with astronomical objectives was launched on October 10, 1946, by Dr. R. Tousey and his colleagues of the U. S. Naval Research Laboratory, and it obtained the first ultraviolet spectrum of the sun in the region between 2,900 and 2,100 Å with a resolution of about 3 Å.

It should be noted here that scientists had been trying to overcome the barrier of the terrestrial atmosphere for many years by using large balloons, which could ascend above most of the atmosphere. One should mention the historic flights of Auguste Piccard who around 1930 managed to reach altitudes in excess of 20 km to obtain better measurements of cosmic rays. In the post war years the balloon technology was developed even further and unmanned balloons were able to reach heights in excess of 40 km leaving below them more than 99 % of the earth's atmosphere.

The first balloon-borne telescope was launched in Europe. A group of English and French scientists obtained photographs of the solar granulation in the late 1956 and early 1957 with a manned balloon carrying an 11-inch refractor, from an altitude of about 9 km. Better known and more impressive were the balloon observations of the Princeton group, headed by Prof. Martin Schwarzschild, which in the late 50's, in the 60's, and in the 70's made important astronomical observations with Stratoscope I and Stratoscope II.

During the 60's many other groups have used balloons and special high altitude jet airplanes to carry their instruments above most of the earth's atmosphere. These experiments included not only optical and infrared observations, but also observations in the hard x-ray and the γ-ray regions of the spectrum. Balloons, airplanes, and rockets were of course the forerunners of the satellites which are the true space vehicles. One should not discard them, however, as antiquated tools of the past because, in spite of their shortcomings, they still have many attractive features, especially their low cost, which make them useful companions of the satellite program of space astronomy.

The first artificial earth satellite, the famous Sputnik I, was launched by the Soviet Union on October 4, 1957. This historic event is considered the actual birthday of the space age because it was the first time that a man-made object, a satellite as it came to be known, was placed in orbit around the earth. The first American satellite, Explorer I, which had a total weight of only 31 lbs., was launched on January 31, 1958. In spite of its very modest payload it made a very important discovery by detecting for the first time the Van Allen belts. This preview was a typical example of the exciting and often totally unexpected discoveries that were to come in space physics and space astronomy.

Figure A.2-I shows the cumulative number of successful satellite launchings by all nations, nearly 1000 of them, up to the end of 1970. It should be noted that some of these launchings placed simultaneously more than one satellite in orbit. Also, they often place several other objects, such as the last stages of their launching rockets, in orbit so that by the end of 1970 there were more than 2,000 man-made objects in orbit around the earth, in spite of the fact that another 2,500 or so had already decayed into the earth's atmosphere. The number of these space debris continues to increase rapidly, and in typically man fashion, we might be turning even outer space into a junkyard.

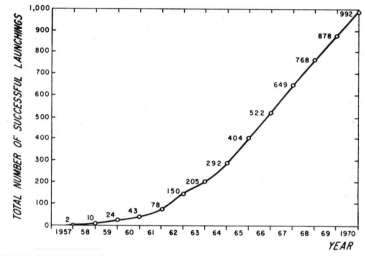

FIGURE A.2-I   Cumulative number of successful satellite launchings by all nations up to the end of 1970

The rate of growth of space technology has been phenomenal. Just hitting the moon with Lunik 2 in September of 1959 was considered a miracle of space navigation and only 10 years later, on July 20, 1969, the first astronauts set foot on the moon after a space trip of unbelievable complexity and precision. It is important to note here that the space program would have been impossible without high-powered computers and radio communications. It is an interesting historical phenomenon that many of these areas of advanced technology were developed independently but almost simultaneously and ultimately by combining together made possible the miracle of the space age.

The moon was the center of attention in the first 12 years of the space program and before it was finally conquered by men it had been repeatedly examined by the Luniks, the Rangers, the Orbiters, and the Surveyors which had sent back more than 100,000 close-up pictures of the lunar terrain and had carefully sampled the soil of the lunar surface. Now scientists have in their laboratories many samples of lunar rocks which were brought back to earth either by the Apollo astronauts or by the automated probes of the Russians.

The study of the terrestrial environment is of great interest and a relatively easier task than space astronomy. For this reason it occupied a major part of all space activity during the first decade. The study of the upper atmosphere, ionosphere, magnetosphere, and interplanetary space have produced a large amount of new knowledge which we have discussed in the first six chapters of this book. There were, of course, also many practical applications of satellites such as *communication satellites* (Telstar, Early Bird, the Russian Molniyas, etc.), *weather satellites* (Tiros, Nimbus, etc.), *navigational satellites*, and *geodetic satellites*, which have also made decisive contributions in the advancement of their respective fields.

The exploration of other planets has been confined up to now to our two closest neighbors, Mars and Venus. The Russians, with their spaceships, Venera 4, 5, 6, and 7 have parachuted scientific payloads in the cloud-covered, scorching-hot atmosphere of Venus which the Americans have also studied with the fly-by missions of Mariners 2 and 5. The Americans have also obtained impressive pictures of the craterpocked surface of Mars with Mariners 4, 6, 7 and 9, while the Russians landed a short-lived probe on it with Mars 3. Now that the moon has been conquered, the exploration of the planets will undoubtably pick up additional momentum by placing orbiting satellites and by soft-landing scientific payloads on both Mars and Venus. Later in the 70's missions to Jupiter are planned, and possibly also to the other outer planets. In the late 70's, taking advantage of a rather rare alignment of the major planets, NASA is planning to send one or more space probes on a *grand tour* of Jupiter, Saturn, Uranus, Neptune, and Pluto. Two

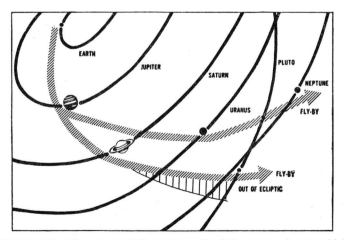

FIGURE A.2-II   The proposed "grand tour" of the outer planets which will take advantage of their favorable alignment around 1980

possible routes are under consideration (Figure A.2-II) which will take 9 and 12 years respectively to reach Neptune.

During the 60's space astronomy emerged finally as an important part of the space program, and is now a strong contender for priorities in the space budget. Going through balloon flights, brief rocket flights, and through sharing space on geophysical or military satellites, space astronomy finally reached independence in the late 60's when large, expensive satellites were placed in orbit especially for astronomical observations. These satellites, carrying a complex array of astronomical instruments, came to be known as *Orbiting Solar Observatories* (OSO's) and *Orbiting Astronomical Observatories* (OAO's). There are even more ambitious plans for the 70's and 80's. These envision ultimately the building in space of a *National Astronomical Space Observatory* (NASO) which will carry a 120-inch *Large Space Telescope* (LST), a 5 to 10 meter x-ray telescope, and many other high quality instruments for γ-ray, x-ray, ultraviolet, and infrared space astronomy.

Man has participated in increasingly daring ventures in space. The first man in space was the Russian cosmonaut Yuri Gagarin who, on April 12, 1961, made the first orbital flight around the earth staying in space for 1.8 hours. The first, and only, woman in space was Valentina Tereshkova, who is now the wife of Andrian Nikolayev, a fellow cosmonaut. The first man to venture outside his spaceship into free space was the Russian Aleksei Leonov on March 19, 1965. This exuberating *space walk*, or *extra vehicular activity* (EVA) as it is officially called, was repeated a few months later by the American astronaut Ed White, who on January 27, 1967, perished together with V. Grissom and R. Chaffee in a flash fire that swept the pressurized cabin of an Apollo spacecraft during a ground test. Three months later, the Russians also had their first fatality when the parachute of their new spacecraft (Soyuz I) failed to open properly, thus causing the death of Colonel V. Komarov.

Figure A.2-III shows the cumulative number of man-days spent in space by both Americans and Russians up to the end of 1970. From this diagram it can be seen that in less than 10 years from the first manned flight we have accumulated more than a year of man-days in space. The setback which the manned program received from the above-mentioned accidents is also clearly seen in the growth curve of this diagram. It was not until October 1968 that both the Soviet Union and the United States resumed manned missions into space.

It is heartening to observe, however, that the initial shock of the accidents was followed by better planning and more hard work, and when man reappeared on the scene, he was wiser, better-prepared, and stubbornly determined the conquer all difficulties. In less than a year from the resumption of manned flights, Apollo 11 landed on the moon and Neil Armstrong was the first man to set foot on the lunar soil. As he was preparing for his first step on the moon, he said that this was "one small step for a man, one giant leap for mankind," and true enough this was the fulfillment of a dream of many generations.

Landing men on the moon probably marks the beginning of a new era when men will explore and later colonize outer space. The question, of course, will always be asked: Is man's presence necessary in the exploration of space or could everything be done equally well, and probably less costly, with unmanned missions? This is a very difficult question to answer, but it might also be an academic one. The reason is that it is totally against man's nature to concede to others and especially to machines all the glory of the new conquests and all the excitement of the new explorations. The

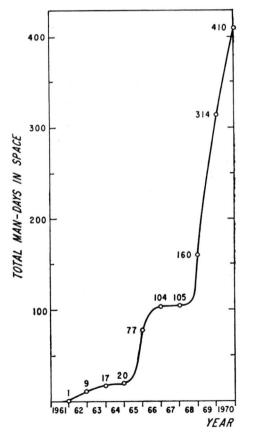

FIGURE A.2-III   Cumulative number of man-days spent in space by both the Americans and the Russians up to the end of 1970

19   Papagiannis

value of man as space explorer is difficult to ascertain in advance. This, however, might be the most crucial decision man ever had to make, because in the long run, it might be directly related to the survival of the human race.

There is little doubt, however, that even if man would be willing to give up his role as space explorer, his presence in space will still be necessary in many future tasks. Skilled space technicians, e.g., will soon be needed to assemble and operate large scientific stations and observatories, either on the moon or in orbit around the earth. These technicians will probably carry out the different experiments, planned and provided by scientists from the ground. The continuous commuting of scientists seems to be a very difficult problem, at least for the near future. On the other hand, the alternative idea of sending a different team of scientists to use a space observatory every few months also seems impractical and uneconomical, because there are not many experiments that could provide optimum use of such a unique facility for several months. As a result the manning of these stations with skillful, specially-trained astrotechnicians, possibily under the general supervision of one or two scientists, seems at the present time the most practical solution.

The first experimental space station, the Russian Salyut, was manned for 23 days in June, 1971, by three Russian cosmonauts. Though the operation of the station was quite successful, the experiment ended in tragedy because all three cosmonauts, Georgi Dobrovolsky, Vladislav Volkov, and Victor Patsayev, were found dead on their return to earth, presumably due to a sudden loss of oxygen in their spaceship, Soyuz 11. Americans hope to place in orbit their first manned space station before 1974. This first American space station will have the name *Skylab* and will be manned by three astronauts for periods of one or more months.

Books for further study and more details on the development of the Space Age can be found in the bibliographies of all the chapters of this book and especially in the bibliography of Chapter 7 and of Chapter 8.

ACKNOWLEDGMENT OF SOURCES

For permission to reproduce the figures indicated below the author wishes to express his thanks as follows

Figure	Author	Publisher	Book or Journal
1.2-II p. 9	Jaccia, 1970	North-Holland Publ. Co.	*Space Research X*
2.6-II p. 59	Chan and Colin, 1969	Inst. Electrical-Electronics Engineers	*Proceedings of IEEE*, Vol. 57
3.3-II p. 75	Hess, 1968	Blaisdell, Publ. Co.	*The Radiation Belt and the Mag- netosphere*
3.4-I p. 78	White, 1970	Gordon and Breach	*Space Physics*
3.4-II p. 79	Hess and Head, 1968	Gordon and Breach	*Introduction to Space Science*
3.4-III p. 80	Hess and Mead, 1968	Gordon and Breach	*Introduction to Space Science*
3.5-V p. 89	Ness, 1969	Amer. Geophysical Union	*Reviews of Geophysics*, Vol. 7
4.3-I p. 97	Tandberg–Hans- sen, 1966	Blaisdell, Publ. Co.	*Solar Activity*
4.4-III p. 108	White, 1970	Gordon and Breach	*Space Physics*
4.6-I p. 118	Hess and Mead, 1968	Gordon and Breach	*Introduction to Space Science*
4.6-II p. 121	Hess and Mead, 1968	Gordon and Breach	*Introduction to Space Science*
5.6-II p. 156	Kerridge, 1970	MacMillan	*Nature*, Vol. 228
7.2-II p. 198	Strauss and Papagiannis, 1971	Univ. of Chicago, Press	*Astrophysical Journal*, Vol. 164
7.2-IV p. 201	Weeks, 1969	Chapman and Hall	*High Energy Astrophysics*
7.3-I p. 204	Goldberg, 1969	W. H. Freeman and Co.	*Scientific American*, June 1969

Figure	Author	Publisher	Book or Journal
7.3-II p. 205	Papagiannis, 1969	Am. Assoc. Adv. Science (AAAS)	*Science*, Vol. 166
7.3-IV p. 208	Linsky and Avrett, 1970	Astronomical Society of the Pacific	*Publications of the Astronomical Society of the Pacific*, Vol. 82
7.4-II p. 212	Hartz, 1969	Pergamon Publ. Co.	*Planetary and Space Science*, Vol. 17
7.4-III p. 213	Fainberg and Stone, 1970	D. Reidel Publ. Co.	*Solar Physics*, Vol. 15
7.4-IV p. 214	Fainberg and Stone, 1970	D. Reidel Publ. Co.	*Solar Physics*, Vol. 15
8.1-I p. 223	Abel, 1969	Holt, Rinehart, and Winston	*Exploration of the Universe*
8.1-II p. 226	Abel, 1969	Holt, Rinehart, and Winston	*Exploration of the Universe*
8.3-I p. 238	Doyle, 1969	U.S. Govern. Printing Office	*A Long Range Program in Space Astronomy* (NASA SP. 213)
8.4-I p. 248	Papagiannis, 1969	Amer. Assoc. Advanc. Science	*Science*, Vol. 166
8.5-I p. 256	Low, 1969	Amer. Assoc. Advanc. Science	*Science*, Vol. 164
8.6-I p. 261	Doyle, 1969	U.S. Govern. Printing Office	*A Long Range Program in Space Astronomy* (NASA SP-213)
3.5-II	Hess and Mead, 1968	Gordon and Breach	*Introduction to Space Science*

INDEX

Aberration effect 145, 153, 201
Ablation correction 152
Absolute magnitude 222
Absorption coefficient 44. 94, 106
Acoustic waves 98, 142
Adiabatic invarient 72
Airglow 50
Albedo 10, 25, 216
Alfvén velocity 86, 143
Alfvén waves 142
All sky camera 178
Alpha particles 78, 149, 171, 176
Ambipolar diffusion 51, 133
Aminoacids 3
Ammonia 3, 20, 24, 226
Andromeda galaxy (M31) 227, 257
Anemopause 90
Angular cyclotron frequency 71
Angular drift velocity of charged particles
   74
Angular plasma frequency 38
Angular resolution 207, 218, 236, 252
Annihilation of matter and antimatter
   232, 233, 258
Antenna arrays 264
Anticoincidence counters 235
Antimatter 231, 258
Ap index 64
Apollo 11 281
Apparent magnitude 221
Apparent velocity 57
Arc-length of dipole field lines 69
Archimedean spiral 145
Argus project 77
Artifical belts of trapped radiation 77
asteroids 24, 25
Astronomical unit 10, 25
Astrotechnicians 282
Attachment coefficient 32, 50, 187
Aurora australis 178
Aurora borealis 178

Aurora polaris 178
Auroral blackout 186
Auroral bulge 180
Auroral electrojet 163, 180
Auroral isophotes 178
Auroral oval 178
Auroral substorms 180
Auroral zone 178
Auroral zone absorption (AZA) 186
Auroras 178
   diffuse patches 180
   high-altitude red arcs (type A) 182
   poleward surge 180
   purplish-red lower border (type B) 182
   quiet arcs 180
   rayet arcs 180
   westward surge 180
Auto-correlation function 58
Azimuthal motion of trapped particles 74,
   167

B and L coordinate system 80
Barium vapor release 165
Baumbach–Allen formula 98
Beamwidth of an antenna 259
Becklin's object 257
Betelgeuse (α-Orionis) 253
Biermann's solar wind theory 86
Big-Bang theory 1, 229
Binary stars 224, 244, 251
Biomass 4
Black body radiation 270
Black holes 230
Blue shift 251
Boltzmann constant 203
Boltzmann equation 203
Bow shock 86
Bragg crystal spectrometer 200, 202
Bremsstrahlung 103, 119, 186, 243, 246
Brightness 221
Brightness spectrum 263

Brightness temperature   103, 275
Butterfly diagram   107

C IV line   248, 249, 250
Calcium plages   106
Capetown anomaly   76
Carbon cycle   92
Carbon dioxide   3, 4, 5, 13, 20, 22, 23, 24
Carbonic rocks   5
Cassiopeia A   56
Celescope project   249
Centaurus A   245
Cen XR-2   237
Cen XR-4   238
Cepheid variables   224, 251
Cerénkov counters   235
Cerénkov radiation   231
Ceres   24, 25
Chapman electron density profile   33
Chapman and Ferraro theory of the mag-
    netopause   83, 84, 86
Chapman layer theory   30, 54
Character figure   34
Charge density   36
Charge exchange collisions   163
Charge transfer   49
Chromosphere   96
Chromospheric emission lines   249
Chromospheric faculae   121
Closed universe   230, 246
Clusters of galaxies   229
Coherent emissions   264
Cold trap   20
Collision frequency   42
Collisional cross-section   42, 132
Collisional damping   215
Comet tails   86
Comets   219
Communication satellites   279
Complex dielectric constant   44
Complex index of refraction   44
Compton effect   234
Compton scattering   193
Conjugate point   49, 181, 187
Conservation of energy   49
Conservation of mass   135, 139
Conservation of momentum   49, 135
Continental drift   3
Convection   13, 93

Convective equilibrium   95
Convective transport   12
Coriolis force   7
Coronagraph   100
Coronal streamers   101, 197, 211
Coronapause   150
Coronascope I and II   208
Coronium   101
Cosmic ray albedo neutron decay
    (CRAND)   81
Cosmic ray showers   171, 231
Cosmic rays   169, 225, 226
    diurnal anistropy   172
    energy spectrum   261
    generation in pulsars   242
    primary   170, 235
    secondary   170, 235
Crab Nebula   148, 225, 241, 251
Critical distance   138, 142
Critical frequency   40, 258
Critical momentum   170
Critical velocity   137, 141
Cross-correlation   57, 58
Cusps   82
Cyclotron frequency   41, 71
Cyclotron radius   71
Cyg XR-2   244
Cygnus A   227, 245

D-region   29, 47, 183
Daily character figure (C)   64
Debye length   129
Debye shielding   43
December anomaly   55
Declination   65
Deep ocean trenches   3
Degenerate matter   224
DeLaval nozzle   139
Deuterium   201, 234
Devonian   4
Dielectronic recombination   102
Diffraction grating   200
Diffuse nebulae   249
Diffusive differentiation   17
Dioxyribonucleic acid (DNA)   3
Dip of the earth's magnetic field   65
Dipole antenna   259
Dipole magnetic field   65
Dipole magnetic moment   64, 65

Dipole moment   43
Directivity of antennas   259
Dissociative recombination   49
Disturbance corpuscular flux (DCF)   162
Disturbance polar (DP)   163
Disturbance ring (DR)   162
Diurnal anomaly   55
Dolomite   5
Doppler broadening   101
Doppler effect   185
Doublet line of neutral oxygen   182
Drift velocity of barium clouds   165
Drift velocity of ionospheric irregularities
    56, 58
Drift velocity of trapped particles   74
Dynamic spectra of radio bursts   121, 122
Dwarf stars   224

E-corona   101
E-region   28, 47
Early Bird   279
Earth   25
Eclipsing binaries   251
Ecliptic   101, 180
Eddington approximation   11, 94, 271
Effective temperature   10, 221, 271
Electron-positron annihilation   233
Electronvolts (relation to wavelength)   196
Electric conductivity of plasma   133
Elliptical galaxies   227
Enzymes   3
Equatorial anomaly   55
Equatorial electrojet   164
Equatorial trough   55
Equivalent current system   165
Equivalent height   40
Escape of atmospheric gases   14
Escape time   18, 19
Escape velocity   16, 25, 138
Exosphere   10, 17
Expanding universe   1, 228, 230
Explorer 1   76, 277
Extra-long base interferometer   226
Extra vehicular activity (EVA)   280
Extraordinary mode   41
Extreme ultraviolet (EUV)   202

F-corona   101, 151
F-layer (F-region)   28

F1-region   29, 48
F2-region   29, 49
f_{min} 186
Faculae   112
Faraday cup   36
Faraday rotation   185
Field lines   68
Filaments   116, 124
Fish Bowl project   77
Flare stars   250, 251
Flares   112
    brightness   113
    importance   113
Flux density of radiation   268
Flux unit of galactic radio emission   243
Fly-by missions   22, 216
Forbidden lines   101, 102, 182
Forbush decrease   149, 171
Forced harmonic oscillator   38
Forecasting of flare events   159, 175
Formaldehyde   226
Fossil fuels   5
Fourier spectroscopy   255
Free-free absorption   262
Free-free emission   103, 119
    coefficient of   200
"Frozen in" field lines   128, 144

Galactic astronomy   221
Galactic center   225, 264
    infrared emission from   257
Galactic evolution   229, 258
Galactic halo (or corona)   227, 264
Galactic magnetic field   226
Galactic plane   225
Galaxy   225
Gamma (γ)   64
Gamma-ray astronomy   193, 231
Gamma-ray background   236, 246
Garden hose effect   145
Gaunt factor   200
Gauss   65
Gegenschein   151
Geiger counter   76, 196, 200, 239
Generalized invarient latitude   81
Geodetic satellites   279
Geomagnetic activity, indices   64
Geomagnetic anomalies   63

Geomagnetic anomaly of the ionosphere 55
Geomagnetic latitude   65
Geomagnetic longitude   65
Geomagnetic poles   65
Geomagnetic storms   161
  gradual commencement   162
  initial phase   162
  main phase   162
  sudden commencement   161
Globular clusters   227
Gradual storm commencement (GSC)   162
Goddard, R.   276
Grand tour   279
Graphite particles   250
Gravitational collapse   224
Gravitational force   134, 139, 154
Gravity gradient stabilization   260
Greenhouse effect   11, 13, 14, 23
  Grey atmospheres   11, 271
  Grey matter   94, 271
Ground events   176
Ground-based astronomy   193
Group index of refraction   40
Group velocity   40
Gyro-frequency   71, 130, 259
Gyro-radius   71, 129

Hα-line   96, 112, 124, 183
H and K lines of calcium   98, 112
H II region   249, 262, 275
Harmonic oscillator   38
Heaviside layer   28
Heavy elements   1
Heliosphere   9, 52
Helium "burning"   223
Helium lines   98
Hertzsprung–Russell diagram   222
  horizontal track   223
Heterosphere   9
Homosphere   8
Hubble constant   228, 230
Hydrodynamic equation   134, 138, 141
Hydrogen "burning"   1, 92, 222
Hydrogen line at 21 cm   226
Hydrogen molecular   226, 250
Hydromagnetic damping   215
Hydrostatic equilibrium   14, 30
Hydroxyl radical (OH)   219, 226, 257

Inclination (dip)   65
Index of refraction in plasmas   40
Inelastic collisions   201
Inferior conjunction   21
Inflation of magnetic field   168
Infrared (near, intermediate, far)   252
Infrared astronomy   193, 251
Infrared emission from galactic center 233, 257
Infrared solar observations   208
Infrared spectroscopy   255
Infrared stars, first systematic catalogue 254
Infrared windows of the earth's atmosphere 252
Inner Van Allen belt   77, 81
Inner zodiacal light   101
Intergalactic plasma   246
Internal gravity waves   187
International Commission on Solar-Terrestrial Physics   188
International polar year, first   178
Interplanetary dust   150
Interplanetary magnetic field   144
  sectors of   147
Interplanetary scintillations   148
Interplanetary space   128
Interstellar dust   222, 250
Interstellar extinction   222, 250
Interstellar reddening   222, 250
Invariant latitude   81
Inverse Compton effect   232, 246
Io   219
Ion-atom interchange   49
Ionization chambers   247
Ionogram   41, 186
Ionopause   90
Ionosphere   10, 28
Ionospheric anomalies   54, 61
Ionospheric irregularities   56, 187
Ionospheric sounder   28, 186
Ionospheric scintillations   56, 149, 258
Ionospheric storms   60, 186
Irregular galaxies   227
Isothermal atmosphere   15

Josephson detector   209, 255
Jupiter   24, 25, 279
  decametric radio bursts   218

Jupiter
  magnetic cavity   90
  magnetic field   82, 192

K-corona   101
K, Kp, and ΣKp indices of magnetic activity   64, 187
Kepler, J.   276
Kiloparsec   222
Kirchhoff's law   270

L-corona   101
L-shells   80
Langmuir probes   36
Large space telescope (LST)   253, 280
Leaky bucket model   81
Light cylinder   242
Light year   222
Limb brightening   103
Limb darkening   95, 274
Limestone   5
Line of force   68
Lithium fluoride (LiF)   240, 247
Local group of galaxies   227
Local thermodynamic equilibrium (LTE)   11, 94, 270
Luminosity   92, 222
Lunik 2   278
Lyman-α 47, 217, 249
Lyot B.   100
Lyman continuum   100, 205, 207, 246

M31   227, 257
M82   257
M87   227, 245
Mach number   144
Magellanic Clouds, large and small   227
Magnetic activity, indices   64
Magnetic bottle   172
Magnetic cavity   84
Magnetic field energy density   115, 143
Magnetic field, equatorial   67
  polar   67
  radial component   66
  tangential component   67
  total   67
Magnetic flux   69
Magnetic induction   65
Magnetic intensity   65

Magnetic moment   72
Magnetic permeability   65
Magnetic polarity of sunspots   108
Magnetic poles   65
Magnetic pressure   109, 150
Magnetic stars   251
Magnetic storms (see geomagnetic storms)
Magnetoacoustic waves   100, 143
Magnetoactive plasma   41, 259
Magnetopause   10, 84
Magnetosheath   10, 87
Magnetosphere   10, 63
Magnetospheric tail   83, 87, 88
Magnitude stellar   221, 252, 253
Main sequence   222, 256
Man-days in space   280
Mariner space probes   22, 23, 88, 155, 279
Mars   23, 25, 90, 279
Mass absorption coefficient   269
Mass emission coefficient   269
Maximum usable frequency (MUF)   41
Maxwellian distribution of velocities   15
Mean free path   130
Mean intensity   267
Megaparsec   222
Mercury   20, 25
Mesopause   6
Mesosphere   6
Metastable states   102
Meteor showers   153
Meteorites   152
Meteoritic complex   150, 156
Meteoroids   151
Meteors   151
Methane   3, 20, 24
Michelson interferometer   255
Micrometeorites   152
Micron   251
Microwave solar bursts   119
Mid-day bite-out   55
Mid-latitude trough   59
Mid-ocean ridge   3
Mie-theory of scattering   155
Milky Way   225
Minimum frequency   186
Mirror oscillations   73
Mirroring points   73
Missing matter   230, 246
Modulation collimator   239

Molniya satellites 279
Moon landing 278, 281
Most probable speed 16
Mt. Palomar telescope 222, 252
μ-Meson counters 171

National astronomical space observatory
    (NASO) 240, 280
Navigational satellites 279
Negative bay 164
Negative ion absorption 94
Negative ions 47
Neptune 24, 25
Neutral pion decay 232
Neutral point 115
Neutral sheet 88, 148
Neutral winds 7, 8, 55, 188
Neutron monitors 171
Neutron stars 224, 225
NGC 1275 245
NGC 4151 253, 257
NGC 7027 257
Nimbus 279
NML Cygnus infrared source 257
NO (nitric oxide) 47
Non-thermal radio emission 275
Novae 224, 244, 251
NP 0532 225
Nuclear transitions 234
Nucleon-antinucleon annihilation 232
Nucleus, galactic 225

Oberth, H. 276
Oblique incidence 40
Oda collimator 239
Oersted 65
Opacity 45, 103, 243, 270
Open universe 230
Optical astronomy 193, 202, 251
  solar in space 207
Optically thick 103, 244, 275
Optical thickness or depth 45, 93, 270
Optically thin 104, 244, 275
Orbiter satellites 278
Orbiting astronomical observatory (OAO),
    217, 219, 247, 280
Orbiting solar observatory (OSO) 202,
    204, 280
Ordinary mode 41

Oscillating universe 230
Ottawa index 106
Outer Van Allen belt 77, 81
Ozone 4, 193, 217
Ozonosphere 6, 20

p-p nuclear interactions 232
Pair production 193, 234
Paleozoic 4
Pallas 24
Parabolic electron density profile 33
Parsec 222
Partition function 203
Pegasus satellites 155
Penumbra, sunspot 105, 175
Perihelion 154
Perseus A (3C 84) 245
Photodetachment 47
Photodetectors 254
Photodissociation 4, 5, 6, 193
Photoelectrons 48
Photoionization 6, 48
Photomultipliers 247
Photopion production 232
Photosphere 93
Photospheric faculae 112
Photospheric granulation 96
Photosynthesis 3, 5
Pinching of magnetic field lines 89, 115,
    117, 120
Pitch angle 71, 72
Plages 112, 124
Planck's law 270, 271
Planetary atmospheres, formation and evo-
    lution 1
Planetary meteorology 218
Planetary nebulae 224, 251, 257
Planetary space astronomy 216
Plasma frequency 36, 38, 259
Plasma sheet 89
Plasma waves 119, 121
Plasmapause 46, 52, 188, 259
Plasmasphere 52, 188
Pluto 25
Pogson's rule 221
Polar blackout 186
Polar cup absorption (PCA) 176, 186
Polar substorms 163, 180
Polar wind 59

Polarizability 43
Population I and II stars 1
Pores 110
Positive bay 164
Positrons 200
Post-burst increase 119
Poynting-Robertson effect 153
Precambrian 3
Prepaleozoic 3
Preplanetary stage 257
Primordial soup 3
Prominences 112, 116
    active or sunspot 116
    quiescent 116
    eruptive 117
    surge 117
Proportional counters 200, 239
Proton event 172, 176, 186
Proton-proton cycle 92
Protonosphere 9, 52
Protostars 257
Protosun 1
Pulsars 169, 192, 225, 241

Quasars 192, 228, 251, 258
    3C 273 228, 245, 258
Quiet sun 104

R-corona 103
R-Monocerotis 257
RR-Lyrae 189, 224, 251
Rad 175
Radar meteors 152
Radiation belts 76
Radiation pressure 84, 154
    definition 268
Radiative equilibrium 11, 94, 271
Radiative transfer 11, 94, 271
    derivation of equation 269
Radioactives isotopes 174
Radio astronomy explorer (RAE) 213, 260
Radio aurora 181
Radio blackout 183
Radio brightness of the solar corona 103
Radio bursts, solar 104, 117, 210
    microwave 119
    type II (slow drift) 121, 210
    type III (fast drift) 120, 210
    type III storms 213

type III exciter velocity 211, 213
type III decay time 214
type IV 123
type U 120
type V 120
Radio bursts of Jupiter 218
Radio flux 104
Radio galaxies 227
Radio interferometer 213, 264
Radio spectroscopy 226
Radius of curvature 69
Ranger satellites 278
Rayleigh-Jeans formula 275
Rayleigh scattering 207, 254
Realtime 204
Recotimbination coefficient 32, 48, 49
Red giants 223
Red shift 228
Red spot of Jupiter 24
Reducing atmosphere 3
Reflection coefficient 40
Reflectivity 10
Relative sunspot number 106
Relaxation time 48
Resolving power of telescopes 195, 207
Resonance and fluorescence scattering 217
Rest mass 71
Retrograde rotation 21
Ribonucleic acid (RNA) 3
Rigidity 170
Ring current 64, 162
    eastward 166
    westward 167
Riometer 45, 176, 183
Rhombic antenna 261

Sagittarius A 257
Sagittarius constellation 225, 238
Saha equation 203
Salyut 282
Satellite launchings 277, 278
Saturn 24, 25
Scale height 15, 99, 129, 187
Schwinger frequency 82
Scintillation (twinkling) 207
Scintillation counters 200, 231, 235, 239
Sco XR-1 237, 243, 251
Sea level events 176

Seasonal anomaly   55
Seeing   207
Seyfert galaxies   227, 251, 257
   NGC 1275   245
   NGC 4151   253, 257
Shell stars   224
Shock waves   99, 120, 144
Short wave fadeouts (SWF)   183
Si IV line   248, 249, 250
Sidereal period   92, 144
Sidereal year   21
Silurian   4
Skylab   282
Small astronomical satellites (SAS)   236
Solar activity   112
Solar constant   10, 160
Solar corona   96
Solar cosmic rays   169, 177, 186
Solar emission (steady, slow varying, and
      transient) components   158, 159
Solar flare effect (SFE)   184
Solar granulation   96
Solar nebula   2
Solar proton albedo neutron decay
      (SPAND)   81
Solar proton events   176
Solar radio bursts (see radio bursts)
Solar system, age   1
Solar system, formation   1
Solar-terrestrial relation   158, 189
   energy considerations   160
Solar wind   86, 160
   typical parameters   149
   velocity   86, 142, 149
Solar x-ray astronomy   196
Solar x-ray bursts (see x-ray bursts)
Somnium   276
Sound velocity   86, 99, 143
Sound waves   99
South Atlantic anomaly   76
Soyuz I   280
Soyuz II   282
Space astronomy   193, 280
   general objectives   195
Space observatory   253
Space radio astronomy   193
   galactic   258
   solar   209
Space technicians   282

Space walk   280
Spallation reactions   201
Spark chamber counters   235, 236
Specific intensity   267
Spectral classes of stars   221
Spectroheliograms   112, 202, 203
Spectroheliograph   202, 204
Specular reflection of x-rays   201
Spicules   97
Spiral arms   225
Spiral galaxy   227
Sporadic-E   56, 187
Spread-F   56, 187
Sputnik I   277
Sq-current   63, 164, 184
Starfish project   77
Statistical weight   203
Steady state theory   231, 233, 246
Stellar wind   250
Stereo   213
Stocke's law   152
Storm commencement (SC)   162
Stratopause   6
Stratoscope I and II   207, 218, 253, 277
Stratosphere   6
Streaming angle   145
Subflares   113
Sudden cosmic noise absorption (SCNA)
      184
Sudden enhancement of atmospherics
      (SEA)   184
Sudden frequency deviation (SFD)   185
Sudden increase of the total electron con-
      tent (SITEC)   185
Sudden ionospheric disturbances (SID)
      60, 159, 183
Sudden phase anomaly (SPA)   184
Sudden storm commencement (SSC)   161
Sunspot groups, types   112
   unipolar (class-α)   111
   bipolar (class-β)   111
   complex (class-γ)   111
Sunspot number   106
Sunspots   105
   following   108, 109
   leading or preceding   108, 109
   magnetic field   111
   penumbra   105
   temperature   110

Sunspots
  umbra 105
Superclusters of galaxies 229
Supergalaxy 229
Supergranulation 96
Supernovae 224, 251
Supersonic flow 138, 142, 144
Surface magnetic anomalies 63
Surveyor satellites 278
Symmetric Bragg diffraction 240
Synchrotron radiation 82, 119, 227, 241, 261, 262
  half-life time 83
  total power 83
Synchrotron self-absorption 264
Synodic period 21, 93

T-Tauri stars 257
Taurus-A 225
Tau XR-1 225, 241
Tail of the magnetosphere 83, 88
Tails of comets (Type I, II) 86
Telemetry 216
Telstar 279
Thermal conductivity 100, 131, 133
Thermal detectors 254
Thermal energy density 114, 143
Thermopause 6
Thermosphere 6
Three-body collisions 49
Three degree ($3°K$) background radiation 192, 230, 233, 258
Tiros 279
Topside sounder 36
Total electron content 185
Transition zone 97
Transmission coefficient 40
Trapped radiation 73
Travelling ionospheric disturbances 58, 187
Tropopause 6
Troposphere 6
Trough, mid-latitude 59
Tsiolkovsky, K. 276
Tube of force 69, 146, 188
Turbidity 195
Turbulence, atmospheric 195, 207, 252
Two-body collisions 49
Two-fluid model 133, 142

Ultraviolet astronomy 193, 202, 246
Umbra, sunspot 105
Universal time (U.T.) 64
Upper ionosphere 51
Uranus 24, 25, 218
Uvicon tube 247, 249

Van Allen belts 77, 277
Variable stars 224, 256
Velocity amplitude 99
Venera space probes 22, 279
Venus 5, 21, 25, 90, 279
Verne, J. 276
Vertical motions of ionization 188
Virgo A (M87) 227, 245
Virgo clusters of galaxies 229
Virgo XR-1 245

Weather satellites 279
Wells, H. G. 276
Whistlers 53, 188
White dwarfs 224
Wien's displacement law 102, 196
Wilson effect 105
Wind motions in the thermosphere 7, 8, 55
Wolf–Rayet stars 224, 251
Wolf sunspot number 106

X-corona 102
X-ray astronomy 193, 237
X-ray bursts, solar 104, 117, 196
  class I (thermal) 119, 197
  class II (non-thermal) 118, 197
X-ray flux 104
X-ray emission from the solar corona 196
X-ray sources, distribution 238
X-ray sources, types 240
X-ray spectroheliograms 202
X-ray telescopes 240

Year, sidereal 21

Zeeman splitting of lines 111, 207
Zodiacal light 101, 150
Zones of trapped radiation 76
Zurich classification of sunspot groups 112
Zurich sunspot number 106